MW01378737

Production Systems Engineering

About the Author

Richard E. Gustavson is the president of Systems Synthesis, Inc., creator and developer of system design methods, which include software packages for unique economic justification procedures, determining assembly sequences, developing assembly process plans, establishing task/resource matrices, and synthesizing cost-effective production systems. He has more than 40 years of industrial and consulting experience in product design and manufacturing.

Production Systems Engineering

Cost and Performance Optimization

Richard E. Gustavson

New York Chicago San Francisco
Lisbon London Madrid Mexico City
Milan New Delhi San Juan
Seoul Singapore Sydney Toronto

The McGraw-Hill Companies

Cataloging-in-Publication Data is on file with the Library of Congress

Production Systems Engineering

1 2 3 4 5 6 7 8 9 0 DOC/DOC 1 9 8 7 6 5 4 3 2 1 0

ISBN 978-0-07-170188-4
MHID 0-07-170188-5

The pages within this book were printed on acid-free paper.

Sponsoring Editor
Taisuke Soda

Acquisitions Coordinator
Michael Mulcahy

Editorial Supervisor
David E. Fogarty

Project Manager
Satvinder Kaur, Aptara, Inc.

Copyeditor
Sunita Dogra, Aptara, Inc.

Proofreader
Shashi Lal Das

Indexer
Ariel O. Tuplano

Production Supervisor
Pamela A. Pelton

Composition
Aptara, Inc.

Art Director, Cover
Jeff Weeks

To the memory of Richard Tompkins Gustavson
My son was an intellectual giant who died while serving the country he loved so much. Although far from his main interests, his keen insights into the kind of problems that I was trying to solve were truly invaluable.

To the memory of Ferdinand Freudenstein
Late professor of mechanical engineering at Columbia University, he is universally known as "the father of modern kinematics." His fundamental idea of using synthesis methods for the solution to an engineering problem has guided my efforts for an entire professional career.

To the memory of Ludwig van Beethoven
A genius so universal that his popularity has never ceased to grow. He was a man of astonishing complexity and overpowering intelligence. His music has uplifted me from occasional discontent and truly inspired me countless times.

To the memory of Leonardo da Vinci
The fifteenth-century genius exemplifies the inventive capacity of mankind better than anyone in recorded history. When I need cerebral inspiration, a review of some aspect of his work generally motivates me.

Contents

List of Tables xi
List of Illustrations xiii
Preface xvii

1 Finding a Better Method for Manufacturing System Design . . . 1
1.1 The Situation . . . 3
1.2 Internal Organization of Companies . . . 4
1.3 Economic Justification . . . 6
1.4 Manufacturing Methodologies . . . 7
1.5 Solving the System Design Problem . . . 10
1.6 Summary . . . 11

2 Results from Initial Studies . . . 13
2.1 Background . . . 15
2.2 Basics of System Design . . . 16
2.3 Available Time for a Resource . . . 18
2.4 Allocation of Time Used . . . 19
2.5 Flexibility of a Resource . . . 19
2.6 Fixed Cost of a Station . . . 20
2.7 Variable Cost for a Task . . . 21
2.8 Quality Rating . . . 22
2.9 Solution Procedure . . . 22
2.10 Input Data . . . 24
2.11 Results . . . 24
2.12 Summary . . . 27

3 Real-World Applications Lead to Enhanced Understanding . . . 29
3.1 Introduction . . . 31
3.2 Fundamental Principles . . . 32
3.3 Using a Component/Mate Schematic . . . 32
3.4 Establishing the Process Plan for an Assembly System . . . 33
3.5 Specifying the Economic Constraints and Production Requirements . . . 41
3.6 Determining a Group of Usable Systems . . . 47
3.7 Details of the Best System . . . 53

3.8 Management Overview of a System 58
3.9 Spectrum of Systems for a Range of Production Volumes 62
3.10 Summary 62

4 Stochastic Analyses Added to Deterministic Results 63
4.1 Introduction 65
4.2 Applicable Discrete Event Distributions 66
4.3 Using the Triangular Distribution 68
4.4 Application to a Manufacturing System 71
4.5 Using the Exponential Distribution 78
4.6 Application to Synthesis of Systems 89
4.7 Summary 92

5 Initial Look at System Configurations 93
5.1 Introduction 95
5.2 Geometric Layouts 95
5.3 Schematic Layout Basis 96
5.4 Linear System Layout 97
5.5 Closed Loop System—Without Spacing 97
5.6 Closed Loop System—With Spacing 98
5.7 "U" Cell System 101
5.8 3-D View of a System 102
5.9 Summary 105

6 Multiple Disparate Products Produced by One System 107
6.1 Introduction 109
6.2 Fundamental Principles 110
6.3 Establishing the Multiple-Product Task/ Resource Matrix 110
6.4 Specifying the Production Requirements . . . 111
6.5 Determining a Group of Usable Systems . . . 114
6.6 Details of the Best Multiple-Product System 118
6.7 Management Overview of a System 123
6.8 Summary 131

7 World-Class Versus Mostly Manual Systems . . . 133
7.1 Introduction 135
7.2 The Constant Value Situation 138
7.3 Nonconstant Yearly Costs 141
7.4 Changes in Yearly Production Volume 142

7.5 Changes in Yearly Costs and Production Volume 143
7.6 Summary 145

Appendices

A Determining Allowable Investment 147
A.1 Introduction 149
A.2 Description of a New Technique 149
A.3 Allowable World-Class Investment 157

B Economic–Technological Synthesis of Systems 159
B.1 Introduction 161
B.2 Basic Ideas 162
B.3 Annualized Cost (or Capital Recovery) Factor 163
B.4 Cost Comparison Equation 163
B.5 Utilization 165
B.6 Applicable Technology Chart 165
B.7 Finding the Least-Cost System 167

C Establishing Task Data for Assembly Systems . . 171
C.1 Introduction 173
C.2 Fundamental Principles 175
C.3 Input Data Requirements 176
C.4 Exploded View of the Assembly 178
C.5 The Base Component 178
C.6 The Exploded View 179
C.7 An Assembly Sequence 180
C.8 In-Process Testing 182
C.9 The Best Assembly Process Plan 182
C.10 Summary 187

D Simultaneous Improvement in Yield and Cycle-Time 189
D.1 Introduction 191
D.2 A Different Approach 193
D.3 Evaluating Production Improvement 194
D.4 Expected Production Output 198
D.5 Expected Costs 199
D.6 Summary 201

E Two Case Study Summaries 203
E.1 Case Study Number 21—Automatic Transmission Final Assembly 205

E.2 Case Study Number 24—Automatic Transmission and Differential Final Assembly 206
E.3 Summary 209

F Advanced System Design Procedure 211
F.1 Introduction 213
F.2 Basic Information 214
F.3 Optimizing Assembly 215
F.4 Design of Assembly Systems 217
F.5 Limitations on Program 220

References 221

Index 223

List of Tables

2.1 Representative Costs for Typical Resource Types 23
2.2 Resource Type Data (Portion) for a Typical Electromechanical Product 25
2.3 Task Time and Cost Assignments for Example System 27
3.1 Portion of an Assembly Plan for a Typical Electromechanical Product 35
3.2 Portion of Initial Resource Type Costing for a Typical Electromechanical Product 39
3.3 Portion of Task/Resource-type MATRIX for a Typical Electromechanical Product 42
3.4 Portion of a Specific Solution for a Typical Electromechanical Product 54
3.5 Portion of a Specific Solution Summary for a Typical Electromechanical Product 55
3.6 Management View for Assembly of a Typical Electromechanical Product 60
4.1 Stochastic Results (Maximum Influence) Using Triangular Distributions 74
4.2 Stochastic Results (Minimum Influence) Using Triangular Distributions 76
4.3 Stochastic Results (Maximum Influence) Using Exponential Distributions 83
4.4 Stochastic Results (Minimum Influence) Using Exponential Distributions 85
6.1a Cost and Performance Data for the Best System (Portion) 119
6.1b Cost and Performance Summary for the Best System (Portion) 120

6.2a Management View of the Best System (Product 1, Portion) 124

6.2b Management View of the Best System (Product 2, Portion) 126

6.2c Management View of the Best System (Product 3) 128

7.1 Example Competing System Specifications—Constant Data 137

7.2 Example Competing System Characteristics—Constant Data 140

7.3 Example Competing System Characteristics—Increasing Costs 142

7.4 Example Competing System Characteristics—Increasing Output 143

7.5 Example Competing System Characteristics—Parabolic Output with Constant Yearly Costs . . 144

7.6 Example Competing System Characteristics—Parabolic Output with Increasing Yearly Costs . 145

A.1 Allowable Investment Cash Flow Table—Simple Case 151

A.2 Allowable Investment Cash Flow Table—Complex Case 153

B.1 Demonstration System Cost and Performance Results 169

E.1 World-Class vs. Mostly Manual Comparison—Case Study 21 207

E.2 World-Class vs. Mostly Manual Comparison—Case Study 24 208

List of Illustrations

2.1 General Cost Versus Production Volume Characteristics 18
3.1 Component/Mate Schematic for an Electromechanical Product 34
3.2 Primary Factors for a Typical Electromechanical Product 37
3.3 Applicability of Resource Types to Tasks for a Typical Electromechanical Product 40
3.4 Choice of Resource Types for a Typical Electromechanical Product 40
3.5 Data, Which Can Be Revised, for Programmable Resource Types 41
3.6 Economic and Available Time Data 43
3.7 Tooling Cost and Count Data 44
3.8 New or Used Equipment Data 45
3.9 Intervening Tasks and Multiple Units on Pallet Data 45
3.10 Schematic Logic Diagram for Advanced System Design Procedure 46
3.11 Selection of Minimum Availability for Advanced System Design Procedure 48
3.12 RATING Weighting Values for Advanced System Design Procedure 49
3.13 RATING Weightings and Desired Values for Advanced System Design Procedure 49
3.14 Example Resource Type, Task, Yearly Production, and Title Data 50
3.15 Portion of the General Solution (Type A) for a Typical Electromechanical Product 51

3.16 Portion of the General Solution (type b) for a Typical Electromechanical Product 52
3.17 Portion of the Specific Solution for a Typical Electromechanical Product 57
3.18 Cost and Utilization Behavior for a Typical Electromechanical Product 59
3.19 Unit Cost and Total Investment vs. Batch size for an Electromechanical Product 62
4.1 Uniform Distribution Behavior 67
4.2 Triangular Distribution Behavior 67
4.3 Exponential Distribution Behavior 68
4.4 Options for Stochastic Analysis 68
4.5 Range Limits for Prescribed Uptime 70
4.6 Cumulative Distribution Function for Any Point in a Triangular Distribution 71
4.7 "Inverse" Cumulative Distribution Function vs. Position in Range 72
4.8 Average Workstation Throughput Time—Triangular Distributions 77
4.9 Stochastic Station Times—Triangular Distributions 77
4.10 Stochastic Time Ratios—Triangular Distributions. 78
4.11 Stochastic Production Ratios—Triangular Distributions. 78
4.12 Range Limits For Prescribed Uptime. 80
4.13 Application of Exponential Distribution. 81
4.14 Cumulative Distribution Function for an Exponential Distribution 81
4.15 "Inverse" Cumulative Distribution Function for an Exponential Distribution 82
4.16 Average Workstation Throughput Time—Exponential Distributions 86
4.17 Stochastic Station Times—Exponential Distributions 86
4.18 Stochastic Time Ratios—Exponential Distributions 87
4.19 Stochastic Production Ratios—Exponential Distributions 87
4.20 Count of Longest Station Times 88
4.21 Favorable Production Requirements 88
4.22 Same Random Number Seed Behavior 90
4.23 Unique Random Number Seeds Behavior 91

5.1 Linear Layout for Example System 97
5.2 Tight Loop System #1 98
5.3 Tight Loop System #2 99
5.4 Tight Loop System #3 100
5.5 Spaced Loop System #1 100
5.6 Spaced Loop System #2 101
5.7 "U" Cell System #1 102
5.8 "U" Cell System #2 103
5.9 Isometric View of Spaced Loop System 103
5.10 Isometric View of "U" Cell System 104
5.11 Perspective View of Spaced Loop System 104
5.12 Perspective View of "U" Cell System 104
6.1 Workstation Time Allocation for Two Products . 113
6.2 Workstation Time Allocation for Three Products 113
6.3 General Solution—Unit Cost vs. Yearly Production vs. Investment for 3 Products 116
6.4 System RATING and Ranking (Two Best Systems) 117
6.5a Schematic Layout for the Best System - Product 1 121
6.5b Schematic Layout for the Best System - Product 2 122
6.5c Schematic Layout for the Best System - Product 3 123
6.6 Summary Effort Required at Each Workstation . 130
7.1 Yearly WC vs. MM Savings Characteristic for a Typical Product 139
7.2 Yearly WC vs. MM Unit Cost Improvement for a Typical Product 140
7.3 Yearly IRoR Behavior for Constant Costs and Output 141
7.4 Yearly IRoR behavior for Constant Costs but Parabolic Output 144
A.1 Plot of Income for Two Simultaneous Conditions—Simple Case 152
A.2 Bar Chart for Three Simultaneous Cash Flows—Simple Case 152
A.3 Plot of Income for Two Simultaneous Conditions—Complex Case 154
A.4 Bar Chart for Three Simultaneous Cash Flows—Complex Case 154
A.5 Interactive Web Page for a Simple Case 155

A.6 Annual Cost vs. Investment for Mostly Manual Systems . . . 156
A.7 Annual Cost vs. Investment for World-Class Systems . . . 157
A.8 Investment Comparison of World Class and Mostly Manual Systems . . . 158
B.1 Typical Unit Cost vs. Production Volume vs. Utilization . . . 164
B.2 Example Applicable Technology Data . . . 166
B.3 System Schematic for Example . . . 168
C.1 Sketch of a Sample Product . . . 175
C.2 Initial Exploded View of Sample Product . . . 180
C.3 Revised Exploded View of Sample Product . . . 181
C.4 Final Assembly Sequence for Sample Product . . 182
C.5 Initial Assembly Process for Sample Product . . 184
C.6 Final Assembly Process for Sample Product . . . 185
C.7 Final Assembly Process Plan for Sample Product . . . 187
D.1 Cycle-Time vs. Unit Produced—Logarithmic . . 192
D.2 Cycle-Time vs. Unit Produced—Linear . . . 192
D.3 Cycle-Time Improvement Behavior . . . 193
D.4 Yield Improvement Behavior . . . 194
D.5 Inverted Yield Improvement Behavior . . . 195
D.6 Typical Economic and Time Data . . . 196
D.7 Yield and Cycle-Time Behavior—General Case . . . 196
D.8 Yield and Cycle-Time Behavior—Specific Case . . . 197
D.9 Aggregate Output Characteristic—Specific Case . . . 198
D.10 Yearly Output and Rework Behavior—Specific Case . . . 199
D.11 Typical Materials and Labor Parameters . . . 200
D.12 Example Cost vs. Output Characteristics . . . 200
E.1 Savings vs. Yearly Production—Case Study 21 . 206
E.2 Savings vs. Yearly Production—Case Study 24 . 208
F.1 Input Data for Limited ACM Example . . . 216
F.2 Manufacturing View (General Solution) . . . 218
F.3 Manufacturing View (Best 3 Shift Solution) . . . 219
F.4 Manufacturing View (Best 2 Shift Solution) . . . 220

Preface

Why can't we do a better job of system design?
Isn't there some way to minimize design AND operating costs?
Can't we find a means to eliminate the need for piecemeal continuous improvement?
What variety can be handled by a single system?
Hasn't someone solved this type of problem?

Business managers are always seeking the most efficient and cost-effective (often termed "optimum" or "optimal") methods for operation of their firm. Manufacturers, in particular, have for more than a century sought the best way of bringing forth their products. Despite having spent significant time and money pursuing this goal and achieving success in many technological areas, the overall system optimization has generally eluded them.

A fundamental imperfection frustrated most efforts—they failed to understand that the technology required for their enterprise *can* be intimately combined with the economic requirements. These two activities, applicable technology and economic justification, have been traditionally assigned to totally separate departments in a company; they rarely cooperated, thus missing a golden opportunity. Each group used independent methods, usually computer assisted, to optimize their own activity. They failed to grasp the potential of bringing both sides of the situation—the technological requirements and the economic requirements—together from the start. While maximum allowable costs have sometimes been specified before a system is designed, knowing whether a system would be accepted economically did not occur until after much work concerning the technology had been performed. A method for simultaneously satisfying both economic and technological requirements was badly needed.

This book fills that knowledge gap by revealing, in significant detail for the first time, a modern method that achieves the ever-elusive ultimate solution. It is based upon 25 years of development using actual industrial applications. A new mindset for business managers, particularly manufacturers, will be opened; the vast majority has no idea about what they have been missing. This book uses the field of assembly system design as the primary example and, while taking it significantly beyond current thinking, establishes a new paradigm for the manufacturing world and other, not yet clearly defined, fields.

Many activities are involved in the creation of an optimum system. Generally, the most cost-effective combination of resources that will perform the required tasks is to be determined. For relatively simple cases, such as the examples typically found in textbooks, an optimum is usually somewhat easy to determine. However, the real world is seldom that elementary. The procedure currently used to find the best solution can require very elaborate analyses in a laborious manner. Until the advent of powerful computers for "everyone," optimal solutions were often not even attempted. Today, very complex simulations are often performed for systems that may not actually be close to optimum economically.

After studying this book, you will understand that an entirely different method from any previously learned can be used in a very expeditious manner to determine optimum systems. The technique does not rely on computerized trial-and-error analysis. Instead of somehow estimating a solution and then analyzing it to see if the requirements have been satisfied, the new method directly finds solutions that automatically meet those conditions. The basic process utilized is synthesis (which the dictionary defines as "the combining of often diverse conceptions into a coherent whole" and/or "deductive reasoning"). Synthesis can be most readily described, for current purposes, as the inverse of analysis; any solution found will satisfy the prescribed requirements. You may be wondering why such a wonderful technique has not been generally taught. Since World War II, educational emphasis for technical subjects in the western world has been on the analytical with very little attention paid to the creative. Synthesis combines the two in interesting ways.

The work involved in creating and developing this book's content was initially aided by input from my Draper Laboratory colleagues, in particular Jim Nevins, Dan Whitney, Tom DeFazio, and Jon Rourke along with Steve Graves at the MIT Sloan School of Management. They, along with numerous MIT graduate students, helped to focus my thinking about the fundamentals of the problem. Many engineers and managers from industry have helped to refine various aspects during all stages of development. Publication of this book would not have been possible without the contributions of Senior Editor Taisuke Soda, Project Manager Satvinder Kaur, Copyeditor Sunita Dogra, Marketing Director Michael McCabe, and Editorial Supervisor David E. Fogarty.

RICHARD E. GUSTAVSON
WARE, MA

CHAPTER 1

Finding a Better Method for Manufacturing System Design

1.1 The Situation

Because of the success of some early applications (1970s and 1980s), well publicized in the trade press, industrial robots appeared to be the answer to many manufacturing prayers. They seldom needed a break (except for preventive maintenance), had reasonable accuracy for many tasks, and possessed good repeatability. Initially capable of relatively simple tasks, "pick and place" robots soon began to be constructed having much greater dexterity and power. Articulated devices of many configurations were soon available; linear, circular, and 3-D types have all been implemented.

While applications that are dangerous to humans (e.g., handling of radioactive materials) are best done by some type of robot, most of the applications within a factory were more for show ("I'm more technologically advanced than my competitors") than for economic reasons. In fact, the main impetus was to simply replace direct labor,* and in some cases fixed automation, with programmable automation (the general term for robots). This desire was implemented under the then prevailing "push" method of production; such a manufacturing arrangement has now been generally replaced by the "pull" method.

In any case, manufacturers (both "Greenfield"† and "Brownfield"‡ types) were clamoring for robots for their factories. Robot producers could not keep up with the early demand. Independent consultants were often called in to help product manufacturers determine which of the available robot types should be used as well as to help in their implementation.

1.2 Internal Organization of Companies

Although generally unrecognized at that time, a very serious dilemma faced most manufacturers: the product design people did not directly talk to the manufacturing system design people. A wall had been allowed to be built up over the years between not just those two groups but often between all departments within an organization. Each group was allowed to create its own fiefdom with virtually everyone, except sometimes top management, not permitted to know the "secrets" by which they conducted their activities. In order to effect a change in the modus operandi, the chain of command up one branch of the organization chart and down another had to be gone through. Not only was this incredibly time-consuming, the end result was often no change or something other than what had been requested.

How could such an organizational structure ever have come about? Almost by default; as companies became larger, it appeared highly desirable to optimize each group's activities. The age of the specialist became very popular after

* While going off-shore might produce lower manufacturing costs, the total cost of getting the product to customers would often be higher. Also, the corporation wanted to retain design and manufacturing capability "in-house."

† "Greenfield" applies to a totally new implementation—product and/or manufacturing system.

‡ "Brownfield" applies to an already-implemented situation that needs to be changed.

World War II. Simultaneously, the analytical method* became the favored method for finding the best way of doing something. Thus, while an organization chart might imply significant possibilities for coordination, it seldom occurred unless ordered from the top. Sometimes, this way of doing business went across divisions of major corporations.

As a case in point, I was involved in the product and manufacturing system redesign for a major automotive component. The original request was to help them "implement robots in the assembly line." Fortunately, the design of the product was not yet "cast in concrete"†; therefore, significant simplifications were possible to be implemented. In-house engineering changes were reasonably straightforward since local management had already "seen the light" about coordinating product design and manufacturing—this was mainly due to the efforts of two strong-willed engineers. All was not rosy, however, since some of the parts to be used in the product had to come from another division of the corporation. Until that time, working-level engineers from the two divisions (of the *same* corporation) had not been allowed to discuss redesign directly—they had to use the chain-of-command in both divisions. An issue that could be resolved in a few minutes, one-on-one, would usually take weeks or sometimes months. Fortunately (partially due to the results of the newly allowed cooperation between the two divisions), the corporation soon adopted the concept of "breaking down the walls between product design and manufacturing."

Implicit in this new cooperation between groups within an organization was the idea that costs were to be minimized. It had been assumed that the cost-estimating department had to exist unto itself—after all, a penny saved on each of a million automobiles was an important amount of money. More significant savings were available by adopting some different methods for deciding how to manufacture

* The analytical method assumes that every problem can be analyzed and thus, eventually, an optimum can be determined by repeated analysis. Many types of mathematics can be involved.

† At some point in time the design of a product must be considered complete—that "cast in concrete" point is when manufacturing can begin. Note that subsequent revisions are possible.

products. In many cases, this meant changing the company's mindset from mass production to lean production (Womack and Jones, 1996). It was still necessary to establish the most cost-effective methods for so doing, however.

1.3 Economic Justification

Initial efforts at economic justification of robots were quite creative. Simply comparing the cost of a programmable device to the cost of the labor to be replaced over a prescribed time period seldom produced the desired justification. Depending upon how many company costs were loaded onto direct labor, any investment could be justified. In fact, the aerospace manufacturing world had, for accounting purposes, loaded practically all costs of the enterprise onto direct labor; thus, replacement of any factory worker (whose hourly cost was thought to be in the hundreds of dollars) by a robot appeared to be easily justified. The rest of the industrial world was not anywhere as liberal in their definition of costs that could be loaded onto direct labor; justification was thus a far more difficult task.

Around that time, activity-based costing* gradually began to become the standard method of accounting for costs in manufacturing. This method, not yet generally implemented in manufacturing organizations, has always been at the heart of my approach to engineering economic analyses (Gustavson, 1983) even before it had an "official" name. By this method in addition to treating only those costs directly associated with a particular manufacturing activity (e.g., final assembly), the actual costs of implementing a particular resource-type† to perform tasks‡ could as well be delineated. The total cost of a robot, or any resource-type, would be significantly more than the hardware cost—engineering, design, installation, and debugging costs would all also occur.

* The accounting scheme that assigns costs based only on what is utilized for the activity, not the labor and equipment hours available, as had been done previously.

† A resource type defines a generic class: manual, fixed automation, and programmable automation.

‡ A task is an activity that MUST be performed, possibly in conjunction with others, by some resource type.

Subsequently in this book, the ratio of total cost to hardware cost will be referred to as the "rho factor."

While manufacturers began to get excited about the possibility of implementing some type of automation in their factories, cost and expected savings data for such projects were generally required to be submitted to a financial department which would either approve the project or not. The techniques of engineering economy (Kurtz, 1984) were quite well known to many engineers, but engineering groups were usually allowed no direct involvement in the cost justification process.

After working with the standard method of economic justification about 50 times, it became obvious to me that the method could be inverted. Instead of specifying a cost for an alternative method along with the expected savings (due to using that alternative versus the current base method) for a time period and determining the internal rate of return (IRoR) or present worth, one could specify the savings stream and minimum attractive rate of return (MARR) for zero present worth and calculate the allowable investment (see App. A).

Even today, there are people who do not believe that such an approach is rational, let alone valuable. It is important to realize that this inversion of the commonly used way of solving a problem is at the heart of the synthesis method. Solutions are sought which *automatically* satisfy the desired, or required, behavior. It is no longer adequate to estimate a solution and then analyze (possibly very extensively) it to determine whether the conditions have been satisfied.

Starting with a request to implement robots, a totally new method for designing a manufacturing system continued to evolve. Combining the economic requirements with the technological capabilities led to a straightforward method for determining the most cost-effective system.

1.4 Manufacturing Methodologies

Traditionally, there has been a sharp division between creating a product and the means by which it is to be produced. For much of the twentieth century, totally manual methods were employed in these processes, particularly those involving assembly. Indeed, many important design decisions were based upon ideas sketched on "the back of an envelope." There is little possibility that all of the important factors involved could be included even if the basics were fundamentally correct.

In the late 1970s and throughout the 1980s, manufacturers of all sizes wanted to know how they could replace their manual production methods with "something better." Large companies, making huge volumes of certain products, had already determined that fixed automation was the answer for their situation. Special case applications of robots and other programmable machines were shown to be technologically valuable but required expanding the normal rules for equipment justification in order to be economically viable. Manufacturing plants of that era were characterized by in-process inventory taking up huge amounts of factory space as well as requiring significant non–revenue-producing costs. It had been done that way for as long as anyone directly involved could remember.

During the 1990s and continuing into today, manufacturers of all sizes want to be "agile" or "lean." A totally new paradigm had to be implemented. This generally means that companies seek to be able to respond to any customer very rapidly while minimizing costs; their production philosophy had to change from "make-to-stock" to "make to order." Such a change required a fundamental rethinking of how production should take place. As pioneered by Toyota (Womack and Jones, 1996), the fundamental technique is to produce only what is required; the "pull" method of manufacturing had to replace the "push" method.

In the "push" method of production, each stage of the manufacturing process works at its maximum rate possible regardless of what may be happening at subsequent steps. The only practical way that scheme could be implemented was to have a usually large buffer between the stages. Such an implementation was the reason for the significant in-process inventory once seen in virtually all substantial manufacturing plants. In order to optimize such configurations, very elaborate mathematical simulation methods were created. Depending upon the complexity of the real system, the model could be very worthwhile but often was capable of only simple approximation to the actual system.

In the "pull" method of production, each stage of the manufacturing process works only when the next stage provides notice that it needs input. Another way of thinking about this is that each step of the process has a customer, either internal or external, to which it responds. Using this scheme, the in-process buffers are minimized—being totally eliminated in many cases. The analysis of such a manufacturing system will be much less complex than the traditional simulation methods. Producing only what is required is a fundamental concept of pull manufacturing. In-process inventory, previously up to thousands of units in a single factory, is minimized. Work is not performed at any area until the next level says it is ready for input. Many clever ways to assist in the accomplishment of this goal have been created:

Just-in Time—Only the appropriate components, at the precise time they are needed, are to be in the manufacturing system.

Kan-Ban—A paper trail that enforces just-in-time by moving with each batch of components. This can be eventually automated using bar codes, RFID tags, computers, etc.

Manufacturing Cells—Combining the requirements for a variety of products so that a set of equipment can produce each of them, as required. This primarily applies to fabrication and sometimes to assembly.

Batch-of-One—Being able to manufacture any combination of items with lot size as small as one.

Continuous Improvement—Finding ways to improve the current processes often by combining and/or eliminating manufacturing activities.

Statistical Process Control—Keeping track of how well the process is doing.

Manufacturing had played "second fiddle" to product design for most of the twentieth century. As a result of the "wall" that separated design from production, many problems occurred when the time came to manufacture such products. While major aspects of the fabrication process (i.e., machining, metal forming) had been investigated and physical devices implemented for many years, the assembly process had very little beyond "rules-of-thumb" to guide it. Ideas for improving the ability to assemble (the "micro" aspect) and for designing cost-effective manufacturing systems (the "macro" aspect) began to evolve in the late 1970s. During this period, many people and companies began to realize that the most rational course of action involved product design and manufacturing system design working together; this activity became known as concurrent engineering or simultaneous engineering; the most recent term, involving the entire enterprise as well as vendors and customers, is known as collaborative engineering. The wall had finally started to be demolished!

1.5 Solving the System Design Problem

One of the primary groups involved in concurrent design activities was the Robotics and Assembly Systems Division of The Charles Stark Draper Laboratory* in Cambridge, MA. The vast majority of the useful knowledge about the micro aspects of assembly (i.e., part mating) currently available came from that group. They also, originally in conjunction with the Sloan School at MIT, developed methods for solving the system design problem. While being funded by The National Science Foundation to establish the basics, the group at Draper Laboratory did consulting work for groups within major international companies. Thus, the ideas and techniques being created and developed had immediate real-world application.

This book explains results of some of the discussions with manufacturing companies that have taken place over the past 25 years interspersed with descriptions of how the

* The Charles Stark Draper Laboratory, Inc., was originally known as the Instrumentation Laboratory at the Massachusetts Institute of Technology. Dr. Draper was the perfector of inertial navigation—a major factor in the ability of NASA to send men to the moon and return them safely to the earth. The laboratory is also the financial sponsor of the Draper Prize awarded biannually for outstanding contributions to technology and society; it is equivalent to a Nobel Prize for engineering.

additional requirements or desirabilities have been added to the fundamental ideas of cost-effective manufacturing.

The system design method to be elucidated requires that sequential tasks must be established. Resource types, with cost and performance characteristics, each capable of performing at least one of those tasks must be defined. The fundamental goal is to determine the most cost-effective combination of resource types for a specified production batch.

1.6 Summary

For the example cited in Sec. 1.2, the significant assembly improvements due to the product redesign allowed the parts to be practically "thrown against a wall and have them end up where they belong."* The product was so easy to assemble that direct labor continued to be used—no robots were implemented! The methods described in ensuing chapters were used to establish the fact that people, with appropriate tools, constituted the most cost-effective system for that application. The best production system for any manufactured goods must be determined using the methods described in this book.

* This is the ideal assembly scenario but, to my knowledge, has never been achieved.

CHAPTER 2

Results from Initial Studies

2.1 Background

As stated in Chap. 1, the macro aspect of product assembly had been overlooked as a subject for serious investigation until the late 1970s. The situation was probably due to the idea that assembly could be done only by people who could be shifted around as needed or by high-speed, specialized machines designed for distinct applications. Flexibility and repeatability became important to many manufacturers once a few robot installations exhibited those characteristics; the generic category for such versatile devices was termed programmable automation. Along with direct labor and fixed automation, there were now three generic resource types to be considered in the design of an assembly (or, more generally, flexible automation) system. Each type had cost, performance, and applicability to specific task(s) attributes.

Because the problem of finding the optimal solution to assembly system design was originally undertaken by an assistant professor and a few graduate students from the

Sloan School of Management at the Massachusetts Institute of Technology, it was initially expected that one of the linear programming techniques then available would readily be applicable. The "textbook" cases initially investigated did reasonably lend themselves to such operations research (OR) methods. However, when the assembly of some actual products was attempted, extensive computer run time was required to find a solution.

The OR program required an initial estimate of the solution; the "goodness" of that opening guess actually had little to do with the time required to find a solution, although that circumstance was not understood until later. In an attempt to speed up the process, I devised a scheme for determining the initial educated guess. Rapidly, we found that the OR linear program almost never improved that initial estimate! The methods used—significantly updated, expanded, and improved over time—are the subject of the rest of this book.

2.2 Basics of System Design

The design of an assembly system will be used as the basis for ensuing descriptions and discussions; you should keep in mind that most of the ideas are potentially applicable to a wide variety of system designs. Available time, as well as applicability, will limit the potentially usable technological choices for each task. The general goal is to find a resource that produces the minimum cost on a task-by-task basis. Increasingly interesting is the fact that resources have the ability to perform more than one task (subject to time constraints) with a specific tool/material handling requirement. When all tasks have been assigned, the resulting system will have a very low cost. No method currently exits for absolutely proving that the results are optimum.

The basic scheme used here is akin to dynamic programming (another OR method), but what resources/tools have been allocated, and still have time available, will be important. Also, a scheme for establishing the potential for use on succeeding tasks by including a flexibility factor was implemented. The solutions were thus found by a heuristic method. Because the minimum cost system is being sought, a "quality rating" was established, which is numerically equivalent to the expected unit cost for performing each task with a prescribed tool/resource combination.

Fundamentally, it is necessary to define the tasks, applicable resource types, and time requirements:

- *Tasks*. A series of activities, which have to be performed in a specified order, must be completed. Three parameters need to be identified:
  1. Task time, i.e., duration of activity
  2. Special equipment cost
  3. Special equipment identifier (here, a numerical value)
- *Resource types*. Three generic categories (manual, programmable, and fixed) can be specified. Six parameters need to be defined:
  1. Symbolic name: M for manual, F for fixed, and P for programmable
  2. Resource type cost, i.e., hardware cost
  3. Uptime expected (also known as efficiency)
  4. Operating/maintenance rate
  5. Maximum stations per worker (inverse of indirect labor required)
  6. "rho Factor," i.e., installed cost/hardware cost
- *Time requirements*
  1. Working days per year
  2. Shifts available per work day
  3. Station-to-station move time (within system)
  4. Production volume(s) per year
- *Economic requirements*
  1. Average loaded labor rate (wages plus benefits)
  2. Annualized cost factor

Early on, it was decided that each of these data categories would have its own data file so that the resource type, time, and economics data files could be reused for any other set of tasks.

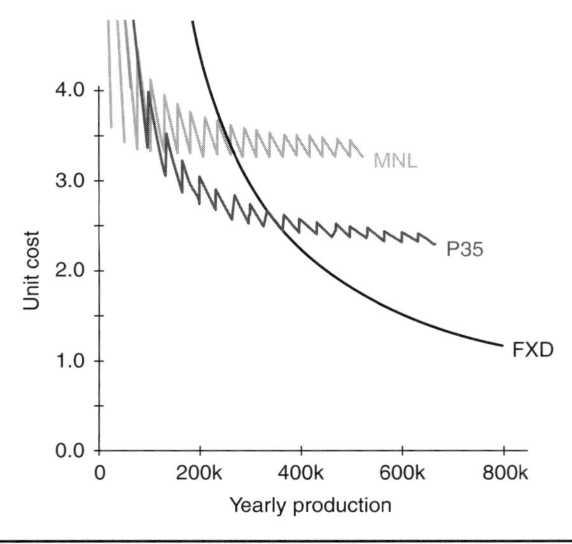

Figure 2.1 General cost versus production volume characteristics.

For each data set (see Fig. 2.1 for an example), it is easy to determine the number of tasks that each resource type can perform as well as the number of tools that would be required.

A ratio

$$n_i = \frac{\text{number of tools}}{\text{number of tasks}}$$

approximates the inverse of the tool changes, which would be required for resource type i. If the expected time for task j is O_{ij} and the tool change time is C_i, then the expected time for resource type i to perform that task can be written as

$$E_{ij} = O_{ij} + n_i C_i \qquad \text{(seconds)}$$

2.3 Available Time for a Resource

A fundamental piece of required information for system design is the time available to perform tasks and necessary tool changes. The maximum time value will be a function of

D = working days per year

S = shifts per day

8 = number of hours in a shift

3600 = seconds per hour

Multiplying these four values together provides the total number of work seconds per year W, or

$$W = 28800 \times S \times D \qquad \text{(seconds/year)}$$

Specification of a yearly production rate Q_y divided into W provides the maximum station time:

$$t_{\text{avl}} = \frac{W}{Q_y} \qquad \text{(seconds/unit)}$$

The actual time available for any resource type must be modified by two characteristics. Except for the special case of all tasks being performed at one physical location, there will be a time increment necessary to move the assembly from station-to-station m_s that must be accounted. Also, each resource type i has an efficiency e_i, also called the uptime expected. Time available is then actually

$$t_{\text{avl}} = \left(\frac{W \times e_i}{Q_y} - m_s\right) \qquad \text{(seconds/unit)}$$

In general, resource types do not have the same uptime factor. Therefore, each type will have a different maximum time available.

2.4 Allocation of Time Used

Each time a task is assigned to a resource type, the available time is reduced by E_{ij} composed of the operation time and a tool change, if required. When the time remaining to be assigned is less than that time, a new resource of that type must be investigated. If no resource type has enough time available, the station must be replicated, i.e., available time must be divided by that same replication number.

2.5 Flexibility of a Resource

When evaluating the capability of a resource type for the group of tasks at hand, some means for determining its flexibility is required. Thus, the flexibility factor for a resource type i may be defined as

$$f_i = \frac{t_{\text{avl}}}{E_{ij}}$$

Variations in f_i will occur only when the production volume Q_y or the task time O_{ij} changes. Initially, the O_{ij} time used (in the preceding formulas) was assumed to be the average time for all tasks that a resource type i can perform. When the operation times are similar, this is a good choice. The flexibility factor is intended to represent the ability of a resource type to perform as many tasks as technologically possible. Thus, the only variable contributing to the flexibility factor is the production volume Q_y. The larger the Q_y is, the lower the f_i becomes, subject to

$$f_i \geq 1$$

Two conditions can produce a $f_i = 1$ value:

1. Number of tools = number of tasks
2. There is no tool change time ($C_i = 0$)

Note that both of these conditions generally apply to fixed automation.

2.6 Fixed Cost of a Station

When the flexibility is greater than 1, the cost of a resource type and/or tooling is temporarily revised to reflect such flexibility. The expected fixed cost for a station is thus defined as

$$F_{ij} = f_{\text{AC}} \rho_i \left[\beta \left(\frac{P_{\text{R}_i}}{f_i} \right) + \gamma P_{\text{T}_{ij}} \right]$$

where f_{AC} = annualized cost factor*
ρ_i = (installed cost/hardware cost) for a resource type (i)†

* Numerically equivalent to the capital recovery factor; it is a function of the minimum attractive rate of return, the capital recovery period, and the residual value (if any).

† This is the "rho factor" for a resource type initially described in Chap. 1.

$\beta = 1$ if a new resource type is required (and $\beta = 0$ if not)
P_{R_i} = hardware price of a resource type i
$\gamma = 1$ if a new tool is required (and $\gamma = 0$ if not)
$P_{T_{ij}}$ = hardware price of tool/material handling for a resource type i to perform a task j

Obviously, the higher the flexibility, the lower the expected fixed cost to perform the task. A special case arises for fixed automation, which is created for only a specific purpose, because

$$P_{R_i} = 0 \qquad \text{and} \qquad \gamma = f_i = 1$$

Comparison of the F_{ij} for each resource type applicable to a particular task leads readily to a selection of the most cost-effective choice. Sometimes, it will be manual, sometimes fixed automation, and sometimes programmable automation. The general behavior is shown in Fig. 2.1 (unit cost decreases as production volume increases). For this particular simplified example, there is clearly a production range for which each of the three generic resource types is most cost-effective. However, such a "textbook" scenario is seldom encountered for actual applications, and a significant rethinking of this concept led to a much-improved solution technique, which is described later in this book.

2.7 Variable Cost for a Task

The variable cost to perform a task (using activity-based costing*) is a combination of average loaded labor rate L_H and the operating/maintenance rate O_i for a resource type. Each resource type also has a specified maximum number of stations per worker M_S, and this number is effectively the inverse of the labor (direct and indirect, i.e., full time equivalent) required for *one* station of that type. Again, the flexibility factor is used in the expected cost calculation to

* This is the modern method for determining costs. See description in Sec. 1.3.

modify the labor rate. This version of the variable cost for a resource type i to perform a task j is expressed as

$$V_{ij} = E_{ij}\left(\frac{L_{\mathrm{H}}}{M_{\mathrm{s}} f_i} + O_i\right)$$

2.8 Quality Rating

Combining the expected fixed cost of a station and the variable cost for a task will provide an expected total cost. This cost is numerically equivalent to the unit cost of a station and is called a quality rating; it applies to only those resource types that are capable of performing a particular task. Minimization of this cost is the primary goal. The quality rating, at this point, is defined as

$$Q_{ij} = \frac{E_{ij} F_{ij}}{W} + V_{ij}$$

For every task that is to be performed, each of the applicable resource types can be compared to all others using this equation. The one with the least Q_{ij} will be selected.

The heuristic scheme described in this chapter will be superseded by the methods in Chap. 3. For now, let's see some interesting characteristics of the system design problem and its solution.

2.9 Solution Procedure

All of the evaluations of task/resource-type data described above were implemented in software, first on a minicomputer and then on a personal computer. The general procedure is as follows:

1. Create/read input data sets. (Table 2.1 exhibits representative data.)
2. Print input data, if desired.
3. Specify yearly production rate(s).

Operation	Time	Tool number	Hardware cost	Annual cost
1	4.00	101	2000	1200
2	5.00	101	2000	1200
3	5.00	102	3500	2100
4	2.00	103	3000	1800
5	5.00	101	2000	1200
6	5.00	101	2000	1200
7	5.00	101	2000	1200
8	2.00	103	3000	1800
9	5.00	101	2000	1200
10	4.00	101	2000	1200
11	3.00	101	2000	1200
12	2.00	103	3000	1800
13	5.00	102	3500	2100
14	5.00	101	2000	1200
15	3.00	104	1000	600
16	10.00	101	2000	1200
17	2.00	101	2000	1200

* Assembly system design:
3.00 s = station-to station move time
0.4000 = annualized cost factor
15.00 = average loaded labor rate ($/h)
225 = working days per year
2.00 = shifts available
Resource data set name: RESOURC4
Task data set name: TASKDAT1
Resource number 1
100 = hardware cost
1.50 = installed cost/hardware cost
60 = annualized cost ($)
85% = uptime expected
0.50 = operating/maintenance rate ($/h)
2.0 s = tool change time
0.83 = maximum stations per worker

Table 2.1 Representative Costs for Typical Resource Types

4. Investigate the flexibility of the resource types.
5. Allocate a resource type to tasks in numerical sequence. (*Note:* Reordering of tasks is NOT allowed!)
   a. Determine the quality rating of the resource type.
      (1) Is tool change necessary?
      (2) Has the tool been already used at the station?
      (3) Has the resource type been already assigned?
      (4) Is time available on that resource type?
   b. Rank the quality rating of the resource types for the task.
   c. Make the resource type and tooling assignments.
      (1) Update all counters.
      (2) Reduce time availability.
      (3) Print time and cost data, if desired.
6. Print statistics for the system, if desired.
   a. Resource(s) used (i.e., time and cost)
   b. Total cost to produce Q_y units.
   c. Capital expenditure for required hardware.
7. Go to step 3, or quit.

2.10 Input Data

An example of typical input data is shown in Table 2.2; annualized cost (or capital recovery) is calculated using

$$K_i = f_{AC} \rho_i P_{R_i}$$

The same factor $f_{AC}\rho_i$ is applied to any tools associated with resource type i. Recall that f_{AC} is the annual cost factor, which is a function of the MARR and the capital recovery period. The other parameters are self-explanatory.

2.11 Results

For a specific yearly production, the maximum available time (ALVTIM) has been determined for each resource type. The

Type	Symbolic name	Resource price	Typical tooling price	Up-time expected	Oper.Main rate	Maximum stations/worker	Installed cost/ hardware cost
Manual	M10	200	2000	85%	0.5	0.833	1.1
	M15						
	M20						
	M25						
Programmable automation	P07	7500	3500	90%	1	8	2
	P15	15000	5000		2	7	2.5
	P35	35000	7500		3	6	3.5
	P45	45000	8500		4	5	4
	P70	70000	12000		7	3	5
Fixed automation	F30	0	30000	95%	1	6	1.5
	F60		60000		2	4	
	F90		90000		3	2	

Table 2.2 Resource Type Data (Portion) for a Typical Electromechanical Product*

maximum number of tasks (NTSKS) and the number of different tools (NTOOL) are calculated as constants. Flexibility (FLXBTY) is determined using the equation given in Sec. 2.5.

700000 units in production batch
Desired production rate is 3.24 units/min
225.0 days required for 2.0 shift operation

	1	2	3	4
AVLTIM	12.7	9.8	9.8	12.7
NTSKS	17.0	13.0	17.0	17.0
NTOOL	4.00	5.0	7.0	13.0
FLXBTY	2.7	1.5	1.3	1.0

Each task must have a resource type assigned to it. There are fixed (annualized) and variable costs associated with that allocation. Depending upon the available time, there will generally be more than one workstation in the resulting system. An easy way to distinguish different stations is to look at the tool number(s)—when a new station is required, the tool number(s) assigned increases by 1000. A particular solution (see Table 2.3) thus has eight stations. Each station may be able to reuse a tool; note that its cost occurs only once. Tool change time is not required for use of the same tool and may be less than the station-to-station move time (as in this case).

The most cost-effective system for this task data, resource type data, and production volume is composed of eight manual stations. Note that the station times are not equal; they rarely will be for real systems. Assigned costs and performance data for this case are as follows:

			Unit cost		Number of	
Resource	Total cost	Number	Fixed	Variable	Tasks	Tools
1	341613	8	0.025	0.463	17	12

3.40 units per minute.
12.0 seconds maximum time at any station.
341613 cost ($) to produce 700000 units.
28800 capital expense ($) for required hardware.

Task	Resource used	Resource cost	Variable cost	Operation time	Tool change	Tool number	Station cost
1	1	60	16994	4.0	0.0	101	1200
2	1	0	21243	5.0	0.0	101	0
3	1	60	21243	5.0	0.0	1102	2100
4	1	0	16994	2.0	2.0	1103	1800
5	1	60	21243	5.0	0.0	2101	1200
6	1	0	21243	5.0	0.0	2101	0
7	1	60	21243	5.0	0.0	3101	1200
8	1	0	16994	2.0	2.0	3103	1800
9	1	60	21243	5.0	0.0	4101	1200
10	1	0	16994	4.0	0.0	4101	0
11	1	0	12746	3.0	0.0	4101	0
12	1	60	8497	2.0	0.0	5103	1800
13	1	0	29740	5.0	2.0	5102	1800
14	1	60	21243	5.0	0.0	6101	1200
15	1	0	21243	3.0	2.0	6104	600
16	1	60	42486	10.0	0.0	7101	1200
17	1	0	8497	2.0	0.0	7101	0

Table 2.3 Task Time and Cost Assignments for Example System

2.12 Summary

A solution method similar in concept to dynamic programming was created for solving the problem of designing assembly systems. At each task, it was assumed that whatever has been selected for prior tasks is optimum and that resource types already allocated (but with time available) cost nothing and that tools already assigned can be used without additional fixed cost. No actual look-ahead is taken, but future events are estimated through the use of the flexibility factor.

Although the techniques described in this chapter are only the beginning of what follows in this book, it is readily apparent that the basic ideas required are straightforward but that the solution to this problem will not be elementary!

CHAPTER 3

Real-World Applications Lead to Enhanced Understanding

3.1 Introduction

Achieving the best possible production systems has usually been, and will continue to be, a difficult and time-consuming trial-and-error process using current methods. Considerable "what–if?" analysis scenarios are gone through with the result that at least one significant constraint is often not properly satisfied. Even manufacturing systems that are considered desirable enough for simulation analysis may not be close to the best available configuration from an economic standpoint. Over the past three decades, a manufacturing system design method that simultaneously satisfies economic and technological constraints has been created and developed; the process used is synthesis (which can be most easily described as the inverse of analysis; any solution found will satisfy the constraints). It is no longer necessary to iteratively estimate a system and then analyze it. The essential idea is to create systems that accomplish as many tasks as possible at each workstation for the minimum activity-based cost. Depending upon many factors, the system will also generally have the best possible line balance.

Because the method can be rapidly utilized, it is valuable at any stage of product or process design. Fundamental economic ideas are discussed in App. A, while basic system design procedures are given in App. B. The techniques have been successfully applied to a wide variety of products at all stages of design.

3.2 Fundamental Principles

Even when the tasks required to manufacture components or to put them together at each level of assembly have been defined, there is still plenty of work to do in order to find the best assembly system. Each task normally has more than one technologically viable method by which it can be accomplished. Although some applicable resource types may be capable of performing only a few of the required tasks, the collection to be used will come from the generic categories of manual, fixed automation, and/or programmable automation. Economic constraints and production requirements also need to be prescribed. This combined group of properties defines the technological and economic limits for the design of a manufacturing system.

There will normally be a number of usable solutions to the system design problem; each solution is a result of specifying the maximum workstation time available. This process can occur automatically, as described later. Ranking the resulting solutions by employing human specifiable *and* alterable criteria readily allows selection of the best system. There is no currently known optimization method for solving a problem as arduous as designing production, or other complex, systems; a pragmatic engineering approach is utilized, which provides very robust solutions. Various tables and graphs provide specifications for any of the synthesized systems.

3.3 Using a Component/Mate Schematic

Depending upon the novelty of the product to be assembled, a means for readily understanding the conditions to be satisfied can be very useful. One of the most practical procedures for starting the process is to create a schematic diagram such as exhibited in Fig. 3.1. Components can be located somewhat like an exploded view of the product (see App. C); major assembly direction is quickly identified. Note that most components should be assembled in the negative z direction such that gravity is beneficial. The type of mate, which must occur between any two components, is specified along with multiples, if required. From this information, an assembly sequence, assembly process plan, and a task/resource matrix can be derived; this procedure may be accomplished manually or aided by automated procedures.

3.4 Establishing the Process Plan for an Assembly System

The first step is to specify the tasks that must be performed. An automated method for determining an assembly process plan is described in App. C. The process plan consists of an ordered set of task descriptions and additional data. Table 3.1 exhibits a portion of a typical *assembly* plan.

In order to establish information about task time and resource applicability, at least the following must be known about each **assembly** task:

Type—the activity that is to take place (14 categories have been identified).

Motions required—linear, planar, or spatial.

Load or force required.

Degree of difficulty—a measure of the complexity of a task (four levels are prescribed).

Task actions—number of activities that must take place (e.g., driving n screws).

In the assembly process plan format used, the tasks will form the rows of a data matrix, while the applicable resource types will constitute the columns.

A method has been developed for establishing the assembly cost and performance characteristics for each of the three generic resource types as well as the additional hardware cost and nominal time for each applicable task (Gustavson, 1990). This activity may be smoothly accomplished

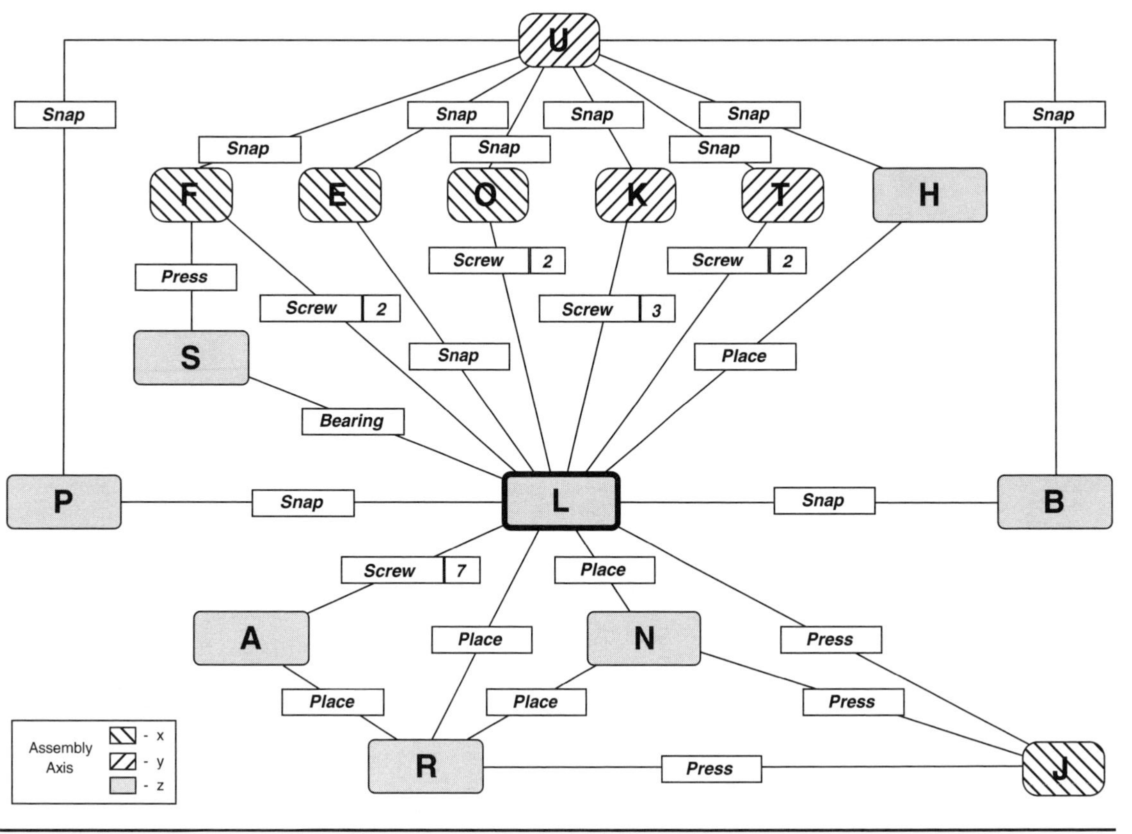

Figure 3.1 Component/mate schematic for an electromechanical product.

Task	Type	Motions required	Load	Degree of difficulty	Task actions	Task description
1	A	Z	4.00	2	1	Attach Al case; Al valve; vacuum element; link (P) to pallet
2	I	Z	4.00	2	1	Install MVH subassy. (B)
3	E	Z	5.00	2	1	Snap fit evaporator case (L) into assembly
4	A	Z	1.00	3	1	Assemble temperature valve (S)
5	T	Y	2.00	2	1	Position temperature valve actuator (F) and tighten fasteners
6	T	Y	1.00	2	1	Position solenoid #1 (O) and tighten fasteners
7	E	Y	1.00	2	1	Snap fit vacuum element #2 (E) into assembly
8	P	Z	0.10	1	1	Place motor-to-case sealant (C) into position
9	P	Z	4.00	1	2	Place motor & fan; Isolator (H) into position
10	E	X	1.00	4	1	Snap fit harness (U) into assembly
11	M	X	0.00	2	1	Test assembled components

Envelope size: X, 15.6; Y, 23.0; Z, 11.5.
Relative assembly difficulty: 1.173.
Task description data set name: ACM-FA-B.

Table 3.1 Portion of an Assembly Plan for a Typical Electromechanical Product

by an experienced manufacturing engineer who can often readily see the need to modify particular data that is created using the automated method described here. A particular robot/end effector/part presentation group may be desired (even though it may not turn out to be cost-effective). Very specialized equipment may be necessary as an option. Manual labor should usually be an alternative because some tasks (especially in assembly) are best performed by a human who generally possesses significant flexibility as well as superior eye–hand coordination.

From many possibilities, only a few resource types will actually be applicable in a particular system. The results described here for determining those resources are based upon heuristic data derived from a wide variety of assembly systems; there is no currently known comparable procedure for fabrication systems. Note that the cost and performance data that is obtained is not likely to be precise for any particular application, but it will certainly be reasonable. Recall that all data is easily alterable.

In general, there will be a single manual labor category and a single fixed automation category; the latter includes a variety of equipment appropriate for individual tasks. Exceptions occur when two or more classes of labor are possible and/or very specific devices such as ovens or automated pallet rotators could be used. Programmable automation (e.g., robots) will often have a few cost categories ranging from moderate (able to do only a few tasks) to very expensive (able to perform most tasks); the count of possible types (usually no more than four) is generally a function of the number of tasks to be performed.

There are two hardware cost categories: one for the resource itself and a second for the "station cost" (which includes all additional equipment necessary for a resource type to perform a single task, e.g., end effectors, part presentation, etc.). A manual workstation will have a resource hardware cost for the bench, chair, light, etc., whereas a fixed automation station has no resource cost (as the term is used here) since each task generally requires a unique piece of equipment and thus has only a "station" cost.

Based upon the task requirements, establishment of the cost and performance data will be required for the

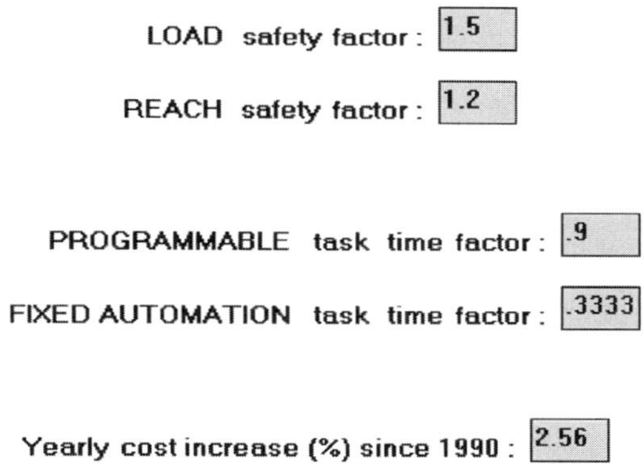

LOAD safety factor : 1.5

REACH safety factor : 1.2

PROGRAMMABLE task time factor : .9

FIXED AUTOMATION task time factor : .3333

Yearly cost increase (%) since 1990 : 2.56

Figure 3.2 Primary factors for a typical electromechanical product.

three generic resource types (manual, fixed automation, and programmable automation). The three fundamental cost-contributing factors are load, reach, and degrees of freedom required; the first two may be altered by specifying a safety factor (see Fig. 3.2).

The basis task time will be manual (where applicable); task time for the other two generic resource types will often be a specified proportion thereof. In the example shown, the nominal time factors assigned are as follows: manual, 1.000; programmable automation, 0.900; and fixed automation, 0.333. Note that task time is the sum of all activities defined using MTM (methods time measurement) or equivalent; it is a function of reach, degree-of-difficulty, number of subtasks (e.g., more than one component to be assembled), and the product relative assembly difficulty.

Load is often the weight of a component, but can be force or torque. The reach requirement is currently derived from the final assembled product envelope; it is assumed that no part will have to be moved farther than from part presentation location to the middle of the product. For current purposes, the degrees of freedom numerical basis will be 4 if linear, 5 if planar, and 6 if spatial. Within each of the three resource types, the load, reach, and degrees of freedom requirements are determined and an estimated cost is assigned to each parameter (based upon heuristic data, when found automatically). By dividing each of the three estimated cost sums by the total of those sums, weighting factors (representing the cost breakdown for each resource type able to perform its applicable tasks) are obtained, which will be applied to the determination of the expected costs as shown in Table 3.2.

The cost for a resource type to accomplish a task is directly related to the expenditure associated with performing all its applicable tasks. Some tasks cannot be performed by every resource type, but at least one must be applicable. The general tests for applicability of a resource type to a task are

Type of task

Degree-of-difficulty involved

Degrees of freedom required

Certain types of tasks require that only a dedicated station will suffice (e.g., oven, automated pallet rotate, automatic test, rework).

When a resource is capable of performing a task, it is necessary to establish

Process time—depends upon mate type and difficulty.

Hardware cost for that resource type.

Tool number—defines a particular set of additional equipment (potentially reusable within a station).

Tool hardware cost for the additional equipment needed.

Using the characteristics for a wide variety of products, it has been found necessary to modify the estimated costs by two multiplying factors: the square root of the task degree-of-difficulty and the specified rate of inflation since the basis year of this model. A variety of such costs can be seen in Table 3.2. While the cost and performance estimates for the manual and fixed automation types will be used as calculated, it is required to reduce the programmable automation types to a few that best represent the process requirements; most companies want to minimize the variety of equipment that they will use. Since it is seldom economical to use programmable automation unless multiple tasks can be performed, a technique is needed for determining which preliminary cost categories offer the highest likelihood of being used in an assembly system. This is accomplished by determining those

		Manual				Programmable				Fixed automation		
Load		32.1%				41.1%				25.7%		
Reach		27.6%				22.1%				9.2%		
DOF		40.3%				36.9%				65.2%		
Time Factor		1.000				0.900				0.333		
Task	**Resource cost**	**Tooling cost**	**Task time**	**Tool number**	**Resource cost**	**Tooling cost**	**Task time**	**Tool number**	**Resource cost**	**Tooling cost**	**Task time**	**Tool number**
1	300	3700	12.0s	1	71000	15500	11.0s	1	0	189000	4.0s	1
2	300	3700	12.0s	2	71000	15500	11.0s	2	0	189000	4.0s	2
3	300	3700	12.0s	3	75500	15500	11.0s	3	0	206000	4.0s	3
4	300	4400	14.5s	4	65500	16000	13.0s	4				
5	300	4200	20.0s	5	72000	15500	18.0s	5	0	291500	7.0s	4
6	300	4200	20.0s	5	64000	14000	18.0s	6	0	276000	7.0s	5
7	300	4200	20.0s	6	64000	14000	18.0s	7	0	276000	7.0s	6
8	300	2600	6.5s	7	30000	8500	6.0s	8	0	90500	2.5s	7
9	300	2600	8.5s	8	51000	11500	8.0s	9	0	134500	3.0s	8
10	300	5300	22.5s	9								
11	300	3800	15.0s	10					0	178500	5.0s	9

2.56% yearly cost inflation factor since 1990.
1.50 LOAD safety factor; 1.20 REACH safety factor.
Task specification data set name: ACM-FA-B.

Table 3.2 Portion of Initial Resource Type Costing for a Typical Electromechanical Product

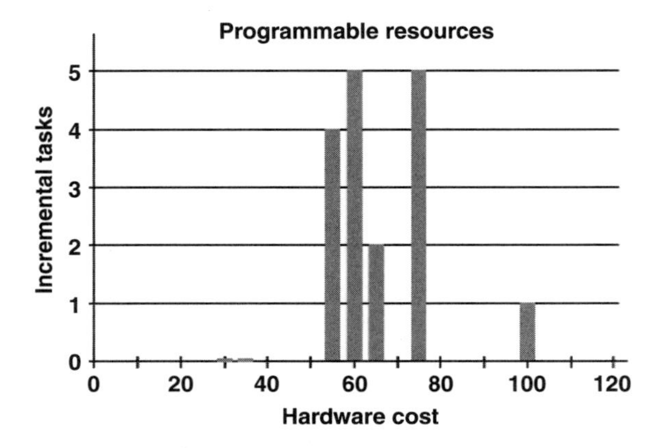

FIGURE 3.3 Applicability of resource types to tasks for a typical electromechanical product.

programmable resource types that have the largest number of consecutive possible task assignments (see Fig. 3.3).

There will be three choices available for the example assembly (a function of the total number of tasks); recall that some tasks cannot be performed by programmable automation. Fig. 3.4 exhibits the choices. Note that two tasks are estimated to require a much lower cost than the minimum selected, but each exists in isolation and is thus not likely to ever be selected as cost-effective. All programmable

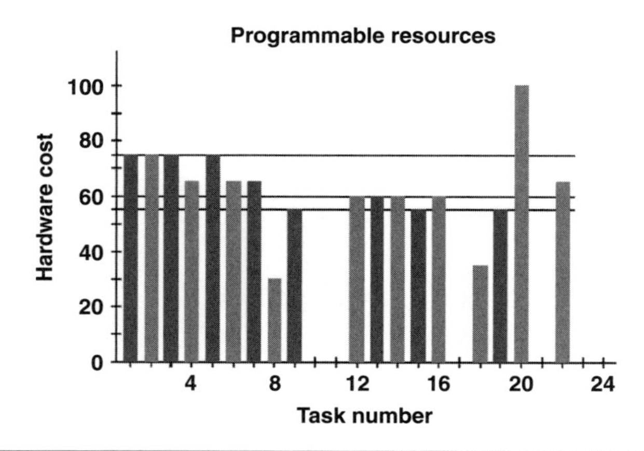

FIGURE 3.4 Choice of resource types for a typical electromechanical product.

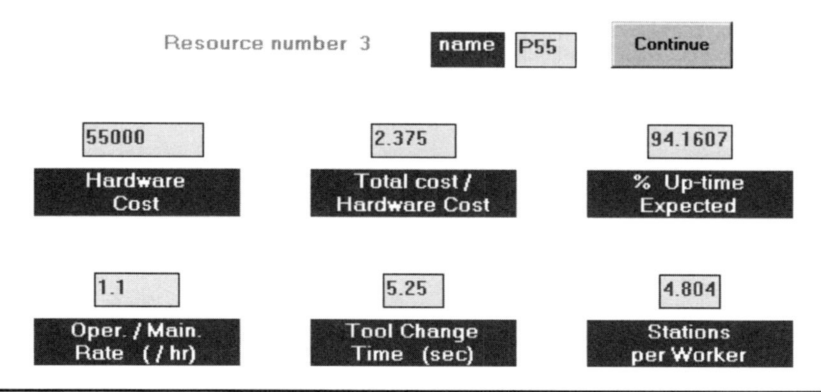

Figure 3.5 Data, which can be revised, for programmable resource types.

automation selections and their cost/performance data can be altered, if desired. Cost and performance characteristics can be estimated from applicable technology (Table 2.2) or, more likely, will be specified (see Fig. 3.5). Any of these parameters can be altered before the data is used in the process of finding the best system to accomplish the required tasks.

The task/resource information that will be needed as economic and technological input for the assembly system design procedure results in the matrix (a portion is shown in Table 3.3). Table 3.3 exhibits example task/resource matrix data; for this case, there are up to *1.4 billion* possible combinations of tasks and resources. Once a production volume is specified, the possibilities are significantly reduced, however. Similar task/resource data would have to be created (manually, currently) for input to the design of a fabrication, or any other type of, system.

3.5 Specifying the Economic Constraints and Production Requirements

Since the goal is to determine the least costly way to do as much work as feasible at each workstation, it is necessary to specify certain parameters that contribute to activity-based costing. Often, no consideration is given to the cost of product materials in the design of an assembly system since that

Resource		MNL	FXD	P55
Hardware cost ($)		200	0	55000
(Total cost)/(hardware cost)		1.20	1.00	2.38
% uptime expected		88.37	94.16	94.16
Operating/maintenance rate ($/h)		0.50	1.17	1.10
Tool change time (s)		2.0	0.0	5.3
Maximum stations per worker		0.83	5.97	4.80
Task	**Description**			
1	Attach Al case; Al valve; vacuum element; link (P) to pallet	12.0s | 101 / 3700	4.0s | 201 / 189000	
2	Install MVH subassy. (B)	12.0s | 102 / 3700	4.0s | 202 / 189000	
3	Snap fit evaporator case (L) into assembly	12.0s | 103 / 3700	4.0s | 203 / 206000	
4	Assemble temperature valve (S)	14.5s | 104 / 4400		
5	Position temperature valve actuator (F) and tighten fasteners	22.2s | 105 / 4200	7.0s | 204 / 291500	
6	Position solenoid #1 (O) and tighten fasteners	20.0s | 105 / 4200	7.0s | 205 / 276000	
7	Snap fit vacuum element #2 (E) into assembly	20.0s | 106 / 4200	7.0s | 206 / 276000	
8	Place motor-to-case sealant (C) into position	6.5s | 107 / 2600	2.5s | 207 / 905000	6.0s | 308 / 8500
9	Place motor & fan; isolator (H) into position	8.5s | 108 / 2600	3.0s | 208 / 134500	8.0s | 309 / 11500
10	Snap fit harness (U) into assembly	22.5s | 109 / 5300		
11	Test assembled components	15.0s | 110 / 3800	5.0s | 209 / 178500	

Table 3.3 Portion of Task/Resource-type MATRIX for a Typical Electromechanical Product

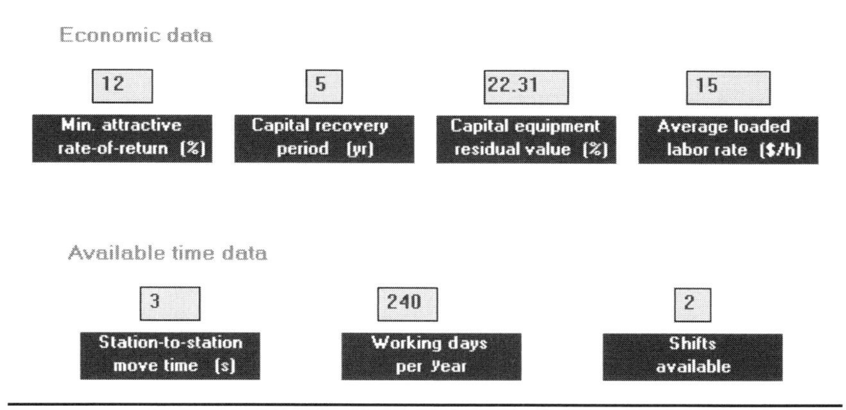

FIGURE 3.6 Economic and available time data.

expense rarely changes due to the process(es) utilized. General and administrative costs as well as any labor overhead (except benefits) will also not be included. What must be considered is as follows:

Capital recovery of the investment—not just depreciation of the hardware. This is numerically equivalent to the annualized cost of the total investment.

Labor rate—including direct and indirect workers (based upon full-time equivalents) in the system.

Operating/maintenance rate—usually a small percentage of the total cost.

The cost parameters that must be specified are (see Fig. 3.6) as follows:

MARR—the corporate minimum attractive rate-of-return for an investment.

Capital recovery period—the time in years over which the investment is to be recovered.

Residual value—calculated as the nondepreciated percentage of the hardware cost, but readily alterable.

Average loaded labor rate—what the nominal worker in the system (direct or indirect) will cost; wages plus benefits only. Note that all labor is charged for a full shift, regardless of time actually required by the manufacturing system.

General limits on the time available at a workstation are derived from the following:

Working days per year—D_y

Maximum shifts available—usually integer (1, 2, or 3), S_d

Station-to-station move time (seconds)—workstation idle time during which no task(s) can be performed, t_m

Units per pallet—almost always one, but can be used to prorate long in–out times (e.g., for an automated guided vehicle)

Production batch size—usually, the units to be produced in a work year, Q_y

Maximum in-parallel stations—usually one, but a higher value could be required; this provides a limit to the number of system solutions (see below) that can be found.

Additional hardware required for a resource type to perform a task has an expenditure called "station cost" here. Some of this cost may be expensed while the rest must be depreciated according to the taxing authority's allowable schedule (7 year MACRS is the default for capital equipment, 3 year MACRS for tooling). Hardware cost is specified by a "tooling cost portion."

The maximum number of tasks that can be accomplished at a workstation is usually determined by the available time but it could be limited by a parameter designated "maximum tools at a station." This condition is generally encountered only when significant time is available at a workstation (see Fig. 3.7).

For resource	Tooling cost portion (%)	Maximum tools at a station
MNL	90	6
FXD	35	1
P55	75	3
P60	75	4
P75	75	6

Figure 3.7 Tooling cost and count data.

For resource	Years already used	Maximum available
MNL	0	299
FXD	0	299
P55	0	299
P60	0	299
P75	0	299

FIGURE 3.8 New or used equipment data.

There is a good possibility that at least some of the equipment to be considered is already being used. Its hardware cost therefore will need to be depreciated by an appropriate amount before being considered as a possibility for inclusion in the new system. For each applicable resource type, the maximum number available for consideration must be specified (see Fig. 3.8). Only new equipment is specified in the current example with large quantities available.

While it is often desirable to allow only consecutive tasks at a workstation (especially for assembly), there are frequently opportunities for "revisits" in flexible manufacturing systems. Seldom used for assembly systems because of the possibility of a person being assigned activity within a robot's work space, nonconsecutive task assignment can be beneficial (e.g., before and after an oven task or an automated testing task). The solution method allows for this possibility through specification of a parameter referred to as the maximum intervening nonassigned tasks as exhibited in Fig. 3.9.

The procedure used can be summarized as shown in Fig. 3.10.

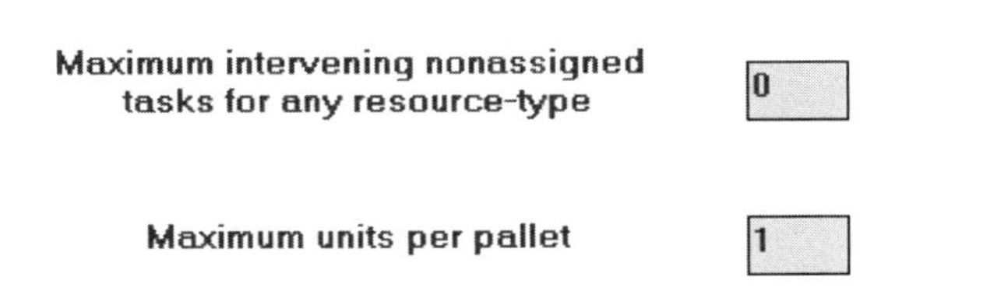

FIGURE 3.9 Intervening tasks and multiple units on pallet data.

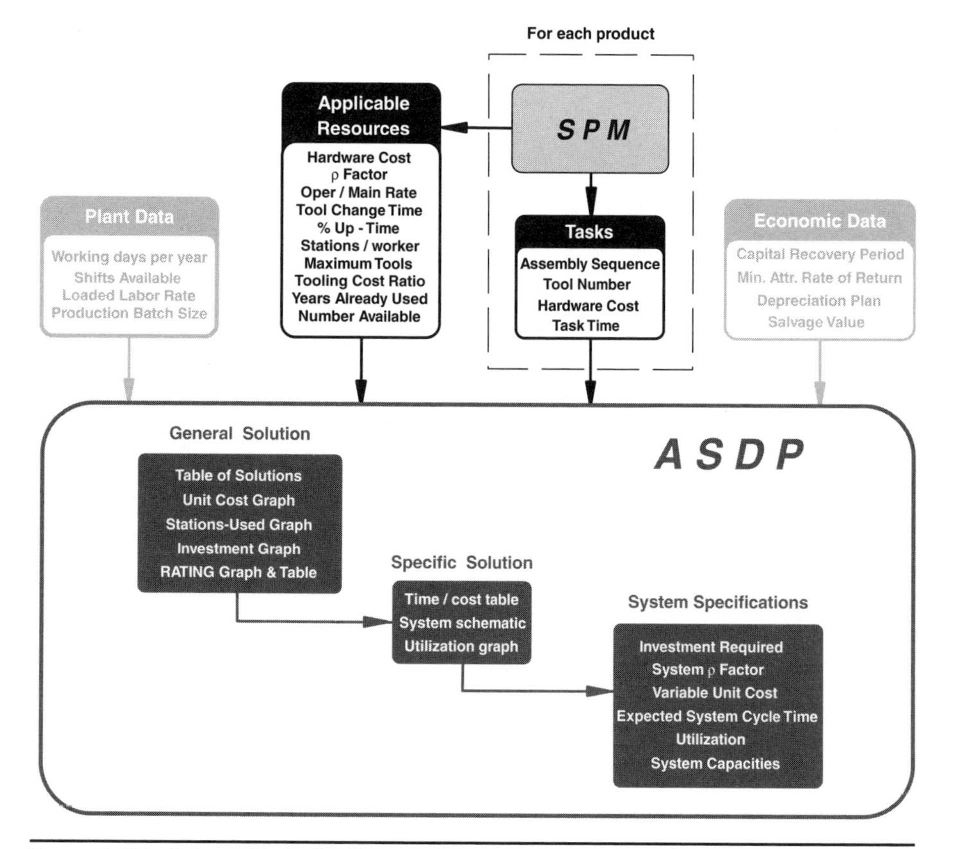

FIGURE 3.10 Schematic logic diagram for advanced system design procedure.

3.6 Determining a Group of Usable Systems

Instead of guessing for a solution to the system design and then analyzing it to see how closely it came to meeting the technological and economic requirements, there is now a direct method (synthesis) to find solutions that always satisfy the imposed constraints. Fundamentally, the goal is to perform as much work as possible at the lowest cost for each workstation. The general solution method is as follows:

1. Calculate the time available at a workstation (for each resource type) from

$$T_{\text{avl}} = \left(\frac{28800 \times S_{\text{d}} \times D_{\text{y}} \times \text{availability}}{Q_{\text{y}}} - t_{\text{m}}\right) \times \text{up - time expected}$$

   where availability is a decimal portion of the total time (maximum = 1.0) and uptime expected is resource type dependent (<1.0).

2. Synthesize the most cost-effective system for that availability.
3. Take an amount of time slightly less than that required for the bottleneck station in the prior system (i.e., reduce the availability).
4. Return to step 1 as long as the in-parallel stations constraint is not exceeded (as defined in Fig. 3.11).

Each of the systems found using this process satisfies all of the technological and economic constraints; in other words, they all are potentially usable systems. Since a method for evaluating each one is needed, every system will be given a RATING relative to all of the others. Three characteristics are given weighting factors; desired values for each of them may be specified:

Unit cost—combination of labor cost, oper./main. cost, and capital recovery of the total investment.

Investment required—installed cost of the system.

Number of stations.

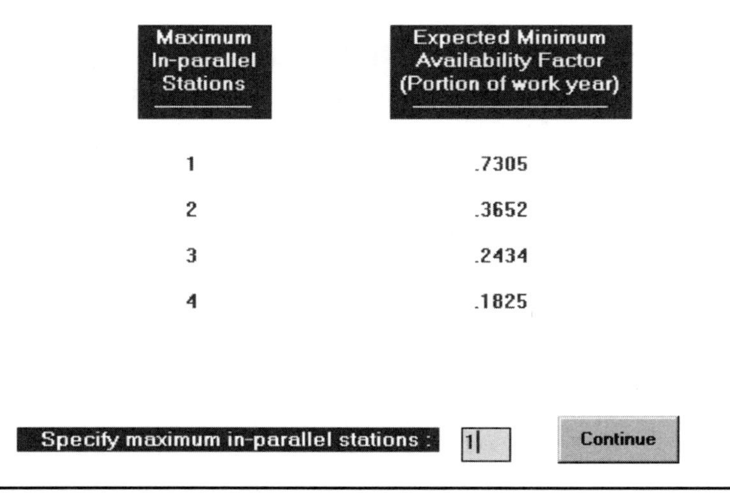

Figure 3.11 Selection of minimum availability for advanced system design procedure.

Since the latter characteristic is usually important only when space is at a premium, the first two often have higher weighting factors assigned. Note that these three weighting factors are expressed as percentages whose sum must be 100. Unless an organization has set particular goals for them, desired values for these characteristics are not often specified. Each solution (system) will have a numerical RATING that provides the basis for a classification of the systems allowing a ranking in descending order; the first solution in the list defines the best system. Although the relative arrangement will not change, the numerical basis for the RATING will be different depending upon the following two:

Specific values not used—maximum availability system (the first one found) is given a RATING of 10 and the three parameters for all other systems are combinatorially compared to its parameters (see Fig. 3.12); all systems will use the prescribed weighting factors. Any system having a RATING larger than 10 is therefore a more desirable system than the initial one. It is possible that no other system is better, especially if all systems are primarily manual.

Parameters specified—the three characteristic values for each solution are compared to those specified (see

The general solution produces a table of synthesized assembly systems which are to be compared.

A subjective RATING uses weights :

Average Unit Cost 40 %

Total Investment 40 %

Number of Stations 20 %

FIGURE 3.12 RATING weighting values for advanced system design procedure.

Fig. 3.13). By using the prescribed weighting factors and a basis of 100, each solution will have a RATING. Any system with a value higher than 100 exceeds the desired set of conditions, although any of the three individual characteristics may *not*. If no system shows a RATING of at least 100, it is easy to determine how near the best system came and thus to determine whether some adjustment in the cost/station requirements or the task/resource data would have to occur in order to improve the situation.

Once the various limitations for the system have been assigned, the resource type data and the task data as well as the production volume must be specified (see Fig. 3.14).

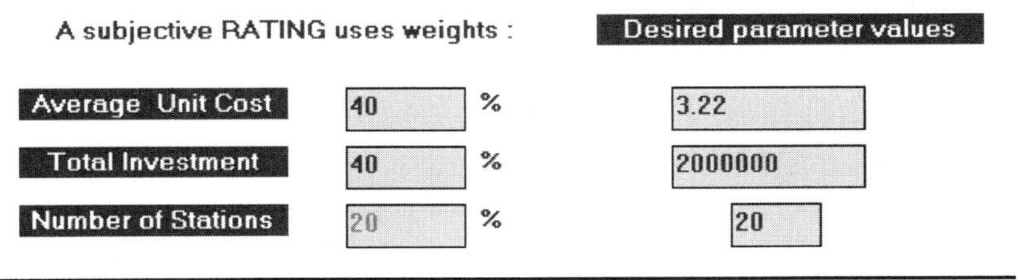

FIGURE 3.13 RATING weightings and desired values for advanced system design procedure.

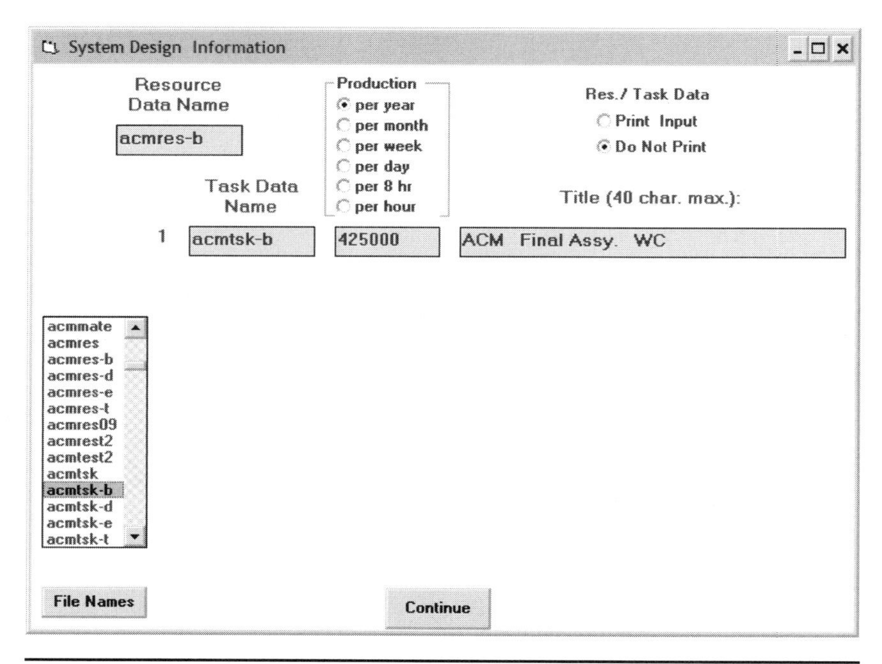

Figure 3.14 Example Resource type, task, yearly production, and title data.

Fig. 3.15 is an example bar-graph and a ranked-listing table showing typical RATING results. The table headings are listed as follows:

Availability factor—the decimal portion of the work year used to determine the upper limit on time available at any workstation.

RATING—the relative merit of a system compared to the others in the solution set or to an ideal.

Minimum pallets—required data for a closed-loop system.

Actual unit cost—sum of the total investment capital recovery, labor charged for full shifts, and system operating/maintenance cost, divided by the yearly production volume.

Total investment—cost of hardware, engineering, design, installation, debugging, etc.

Resource-types data : **acmres-b** Tasks data : **acmtsk-b**

Weighting factors:		
	Unit cost	40.00%
	Investment	40.00%
	Stations	20.00%

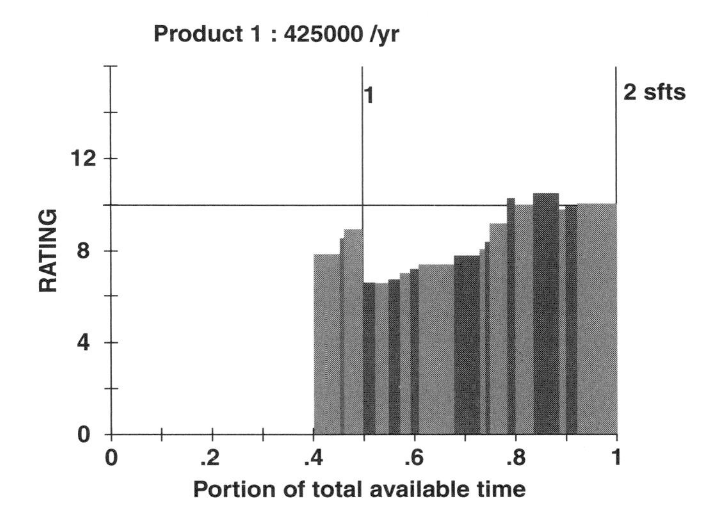

Avail. factor	RATING	Minimum pallets	Actual unit cost	Total (k) Investment	Resources used: MNL	FXD	P55	P60	P75
0.8870	10.48	16	3.155	2039	10 ^	7	1	1	0
0.8001	10.26	16	3.287	2012	11	8 ^	0	1	0
1.0000	10.00	16	3.075	2350	8 ^	8	2	1	0
0.8349	9.98	16	3.257	2173	10 ^	8	1	1	0

Figure 3.15 Portion of the general solution (type a) for a typical electromechanical product.

Resources used—a listing of the resource types, which can be used with the number actually required in each system. The bottleneck station type is indicated by the upward arrow beneath one of the types; at least one station of that type restricts the throughput of the system.

Resource-types data : **acmres-b** Tasks data : **acmtsk-b**

	Weighting factor	Specified value
Unit cost	40.00%	3.22
Investment	40.00%	2000000
Stations	20.00%	20.00

Product 1 : 425000 /yr

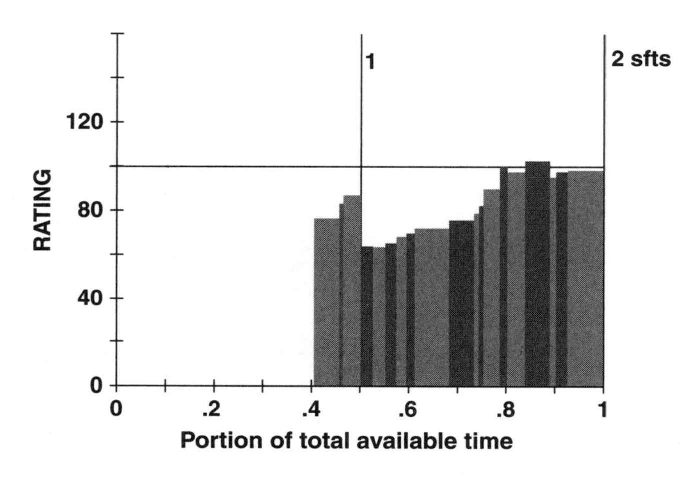

Avail. factor	RATING	Minimum pallets	Actual unit cost	Total (k) investment	Resources used: MNL	FXD	P55	P60	P75
0.8870	101.07	16	3.155	2039	10 ^	7	1	1	0
0.8001	98.94	16	3.287	2012	11	8 ^	0	1	0
1.0000	96.48	16	3.075	2350	8 ^	8	2	1	0
0.8349	96.29	16	3.257	2173	10 ^	8	1	1	0

Figure 3.16 Portion of the general solution (type b) for a typical electromechanical product.

When values are given to the three parameters, the general solution result is of the type exhibited in Fig. 3.16. The ranked listing of solutions is one of the most valuable results of this system design process. Although the better solutions tend to occur near the shift upper boundaries for systems that contain direct labor, this is often not true for primarily automated systems. Whether the highest ranked solution is to be used can be readily decided. For example, the best solution in the table above has at least one MNL (manual) station as its bottleneck. In some company situations, that may not be desirable (or allowable). If such were the case, three of the top four systems shown would not be available for selection as the best system.

3.7 Details of the Best System

The following particulars apply to any of the systems found using the general solution procedure. Although the best solution is readily identified, any availability factor may be arbitrarily selected to use in finding these details. Table 3.4 is an example cost and performance assignment listing. After reiterating the general constraints, a table of characteristics for each available resource type lists

Available time—maximum time available at a workstation

Labor factor/shift—used in determination of labor cost

Oper./main. factor—used in calculating operating/ maintenance cost

Cost and performance details of the synthesized system are then accounted:

Task number—always shown in sequential order (task reordering can occur only by changing the data). More than one task may be assignable at a workstation, depending upon the available time.

Resource used—the resource type found to be most cost-effective for the task(s) assigned. The dash number shows how many of that resource type have been assigned to that point.

Resource cost—the annualized cost (or, capital recovery) for that resource. It is allocated only for the first task at a workstation.

Variable cost—the yearly labor and operating/ maintenance cost. Labor is charged only at the first task at the workstation.

*Resource Types [**ACMRES-B**]*					
	MNL	**FXD**	**P55**	**P60**	**P75**
Hrdwe Cost	200	0	55000	60000	75000
rho Factor	1.20	1.00	2.38	2.50	2.70
Up-time %	88.37	94.16	94.16	94.16	94.15
OprMntRate	0.50	1.17	1.10	1.20	1.40
ToolChange	2.00	0.00	5.25	5.50	6.50
Sttn/Wrkr	0.83	5.00	4.80	4.62	4.27
ToolCost %	90.00	35.00	75.00	75.00	75.00
Max Tools	6	1	3	4	6
Years Used	0	0	0	0	0
No Avlble	299	299	299	299	299
Annl Cost	60	0	34381	39504	53541
OprMntFctr	6363	2897	3082	2825	2421
LbrF/Shift	0.0816	0.0136	0.0141	0.0147	0.0159
*Product 1 [**ACMTSK-B**]; 425000 units/yr.*					
Avltime	22.5	24.2	24.2	24.2	24.2

Synthesized System							
Task	**Resource used**	**Resource cost**	**Variable cost**	**Operation time**	**Tool change**	**Tool number**	**Support cost**
1-1	FXD-1	0	12547	4.0	0.0	201-1	48116
1-2	FXD-2	0	12547	4.0	0.0	1202-1	48116
1-3	FXD-3	0	12547	4.0	0.0	2203-1	52444
1-4	MNL-1	60	70567	14.5	0.0	104-1	1449
1-5	MNL-2	60	70934	20.0	0.0	1105-1	1383

Unused resource available time scale factor 0.887.
212.9 days for 2.0 shifts; 240 days for 1.77 shifts.
12.00% minimum attractive rate of return; 5-year capital recovery period.
22.31% capital equipment residual value; 0.00% tooling residual value.
15.00 $/h loaded labor rate; 3.00 s station-to-station move time.

TABLE 3.4 Portion of a Specific Solution for a Typical Electromechanical Product

Operation time—the task time specified in the data set for the allocated resource type.

Tool change—the time required to change a tool. Since tool change can overlap station-to-station move time, only the difference will be shown (if ≥ 0).

Tool number—allows for the possibility of having multiple use additional support equipment.

Support cost—the annualized cost (or, capital recovery) for the additional equipment (gripper, part presentation, etc.).

The workstation performance data is followed by a report of the system characteristics (see Table 3.5). This summary is composed of the following:

Resource—lists resource types used in the system. For each resource type

- **total cost**—the annualized cost (capital recovery, labor, oper./main.).
- **number used**—how many are in the system.

Time used—the longest expected cycle-time at one of those workstations. This is the sum of operating times

Resource	Total cost	Number used	Time used	Unit cost		Number of		
				Fixed	Variable	Tasks	Tools	Workers
MNL	729641	10	27.2	0.052	1.665	12	11	12.05
FXD	496604	7	25.5	0.954	0.214	7	7	1.40
P55	62789	1	26.0	0.112	0.036	2	2	0.21
P60	64876	1	18.6	0.117	0.036	1	1	0.22
Total				**1.235**	**1.951**			**13.87**

*Minimum pallets in system = 16.
132.56 units per hour.
27.2 seconds cycle-time expected.
88.37% bottleneck station uptime expected.
System capacity: bottleneck, 509011.
Triangular Dist.: RN best, 508053; RN worst, 477417.
Exponential Dist.: RN best, 507719; RN worst, 454700.
15.49/h system operating/maintenance rate.
1353910 cost to produce 425000 units, with unit cost 3.186.
System annualized charge factor 0.2573.
2039280 total investment required.
1809900 for required hardware.
1167690 capital equipment.
642210 tooling.

Table 3.5 Portion of a Specific Solution Summary for a Typical Electromechanical Product*

plus tool changes plus in–out time divided by the uptime expected (expressed as a decimal value).

Fixed unit cost—yearly annualized cost divided by the yearly production volume.

Variable unit cost—yearly labor plus oper./main. cost divided by the yearly production volume.

Number of tasks—total tasks assigned.

Number of tools—total support equipment groups required.

Number of workers—sum of direct (manual station only) and indirect workers.

The performance section is composed of the following:

Units per hour—clock time divided by maximum expected workstation time.

Cycle-time expected—the longest total time at any workstation.

Bottleneck station uptime expected—the value specified in the resource data for the station type that requires the longest station time.

Production capacity of this system—divide work year time by the cycle-time expected. (The triangular and exponential distribution values are explained in Chap. 4.)

System operating/maintenance rate—sum of values for the workstations.

Unit cost—divide total annualized cost by yearly production volume.

System annualized charge factor—a function of the MARR, capital recovery period, and the ratio of total cost to hardware cost for the resources in the system. The lower this value, the easier it would be to justify automation.

Total investment required—hardware plus engineering, installation, debugging, etc.

Total hardware cost

- Total for capital equipment.
- Total for tooling (support equipment).

Resource-types data: **acmres-b** Tasks data: **acmtsk-b**

27.16 seconds Usable Cycle Time
88.37 % Bottleneck up-time 132.56 Units/hr expected

System Capacity: bottleneck 509011
Triangular Dist.: RN best 508055 RN worst 477409
Exponential Dist.: RN best 507723 RN worst 454680

2039280 ($) Total Investment, rho Factor = 1.13
1167690 ($) Capital Equipment 642210 ($) Tooling

13.65 Workers at 15.00 $/hr required
15.49 $/hr System Operating/Maintenance Rate

0.835 Year required for 2.0 Shift Operation
240 Days required for 1.67 Shift Operation

425000 units/yr **$3.155 Each** **0.887 AF**

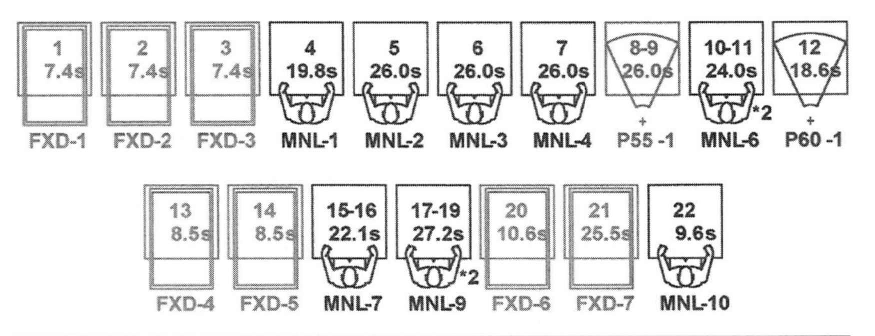

FIGURE 3.17 Portion of the specific solution for a typical electromechanical product.

It is often highly desirable to see a schematic layout for any manufacturing system such as that shown in Fig. 3.17. The three generic types of resources have a specific pictorial:

Manual—bird's eye view of seated person working.

Fixed automation—double *concentric* rectangles.

Programmable automation—simplified bird's eye view of robot workspace.

No other resource type currently has a pictorial representation. The schematic depiction for each workstation contains the following information:

Task(s) to be performed.

Expected cycle-time—includes de-rated operation(s), tool change(s), and in–out.

Resource type specification with system usage counter.

For the example shown, observe that station 14 (MNL-9) is the bottleneck (actually, two parallel stations) and that station 5 (MNL-2), station 6 (MNL-3), station 7 (MNL-4), and station 8 (P55-1) have nearly the same expected cycle-time. A color for each station can represent the highest degree-of-difficulty of a task to be performed there, but is not required:

Blue—easy

Green—moderate

Yellow–brown—complex

Red—must be done manually

Note that the bottleneck stations 14 (MNL-8 and MNL-9) have at least one complex task. It will probably be advantageous to put the best workers at those locations. Having a number of stations with essentially the same cycle-time is quite unusual. However, if trained to line balance, you will be pleased with such a system. Recall that the goal here is to minimize cost but that nearly the same cycle-time for each station *may* also occur.

Sensitivity of cost for variations in production volume for any particular system is easily established. For example, Fig. 3.18 exhibits the rate of change (slope of curve) in unit cost at the prescribed production volume as well as the system utilization. When it is desired to evaluate a production volume that is more than a few percent different, the whole system design process should be repeated using that new manufacturing requirement. Most systems are the best for only a very limited range (plus or minus a few percent) of specific production volumes.

3.8 Management Overview of a System

This view of the system requires cost and performance data from the synthesized system as well as the task descriptions

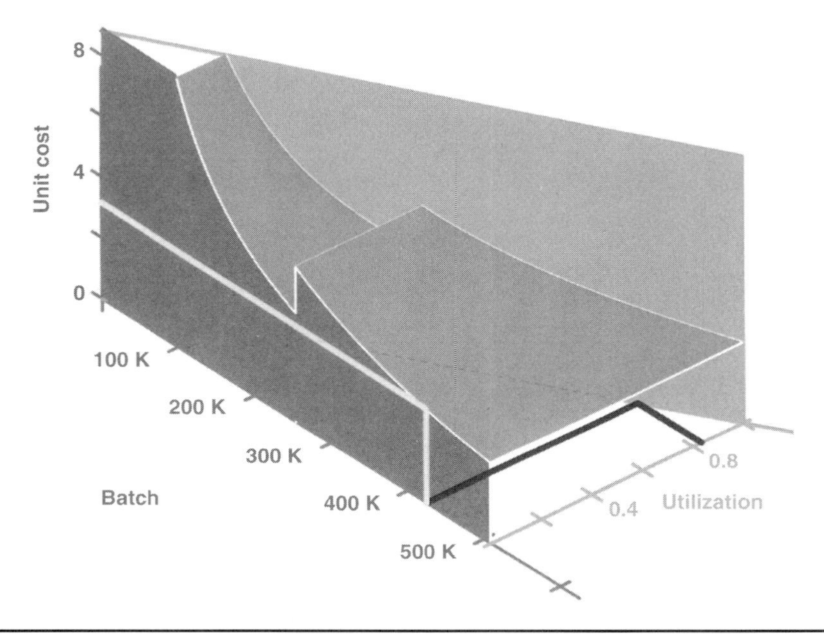

FIGURE 3.18 Cost and utilization behavior for a typical electromechanical product.

from the assembly process plan. After specifying a workstation numbering increment, data (as shown in Table 3.6) can be obtained. Each workstation has the following parameters:

Station number—based upon a specified increment.

Resource type allocated with system usage counter.

Task description—color-coded display is possible.

Effective capacity—units cycled through per hour.

Capital equipment cost—purchase cost.

Tooling cost—includes all hardware that is not capitalized.

Total hardware cost.

Careful inspection of this data provides a sound basis for management to ask for changes in the characteristics of the system. This may lead to higher costs, but at least an organization will now be able to rapidly show how and why that occurred.

Station number	Station type	Description	Eff. cap. unit/h	Captl eq. cost	Tooling cost	Total cost
100	FXD-1	Attach AT case; Al valve; vacuum element; link (P) to pallet	484	123 k$	66 k$	189 k$
200	FXD-2	Install MVH subassy. (B)	484	123 k$	66 k$	189 k$
300	FXD-3	Snap fit evaporator case (L) into assembly	484	134 k$	72 k$	206 k$
400	MNL-1	Assemble temperature valve (S)	181	1 k$	4 k$	5 k$
500	MNL-2	Position temperature valve actuator (F) and tighten fasteners	138	1 k$	4 k$	4 k$
600	MNL-3	Position solenoid #1 (O) and tighten fasteners	138	1 k$	4 k$	4 k$
700	MNL-4	Snap fit vacuum element #2 (E) into assembly	138	1 k$	4 k$	4 k$
800	P55-1	Place motor-to-case sealant (C) into position	138	57 k$	6 k$	64 k$
		Place motor & fan isolator (H) into position		3 k$	9 k$	12 k$
900	MNL-6*2	Snap fit harness (U) into assemb1y	149	1 k$	10 k$	11 k$
		Test assembled components		1 k$	7 k$	8 k$
1000	P60-I	Orient 3 stud seals (M)	193	64 k$	11 k$	75 k$
1100	FXD-4	Position resistor assy. (K) and tighten fasteners	423	121 k$	65 k$	187 k$

1200	FXD-5	Position solenoid assy. (T) and tighten fasteners	423	121 k$	65 k$	187 k$
1300	MNL-7	Place evaporator core subAssy. (N) into position	163	0 k$	2 k$	3 k$
		Place heater core; heater core shroud; clamp (R) into position		0 k$	0 k$	0 k$
1400	MNL-9*2	ASSEMBLE pipe seal (J)	132	1 k$	10 k$	11 k$
		Place cover-to- case sealant (C) into position ALIGN cover (A).		1 k$ 1 k$	5 k$ 7 k$	5 k$ 7 k$
1500	FXD-6	Torque bolts	338	162 k$	87 k$	249 k$
1600	FXD-7	Perform final test	141	252 k$	135 k$	387 k$
1700	MNL-10	Pack/unload assembly	374	1 k$	4 k$	4 k$
Total				**1168 k$**	**642 k$**	**1810 k$**

Required annual capacity = 425000 for Product 1.
Resource types data: ACMRES-B; tasks data: ACMTSK-B

Table 3.6 Management View for Assembly of a Typical Electromechanical Product

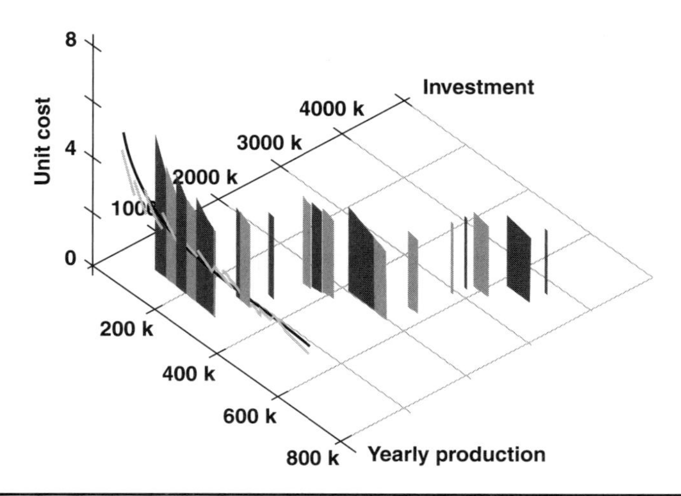

Figure 3.19 Unit cost and total investment vs. batch size for an electromechanical product.

3.9 Spectrum of Systems for a Range of Production Volumes

Now that the procedure for a specific production requirement has been described, let's look at a range of possibilities. Fig. 3.19 exhibits how unit cost varies with production volume as well as the corresponding total investment. Each system section of the spectrum is determined using exactly the same general solution method for the particular yearly production already examined.

3.10 Summary

Starting with a description of the assembly tasks to be performed, you can now get significant assistance in determining the assembly cost and performance characteristics for types of applicable resources. For any system where tasks and applicable resource types can be specified, it is now straightforward to determine a number of usable systems from which the best can be readily established. All of the necessary details are available in graphical or tabular form.

CHAPTER 4

Stochastic Analyses Added to Deterministic Results

4.1 Introduction

For many years, I sought a method that will help people perform a stochastic (i.e., variable time values, occurring randomly) analysis of production systems. Part of the reason for this work is to develop a bridge between the results using the ASDP method (which provides single-valued results) and various simulation programs. Early on, I decided that total activity time (including in/out) at a station would be the basis for analysis. Since the theoretical and expected station times can be determined, it might be straightforward to ascertain how the stations interact as portions of a system. The stochastic throughput results could be quite different from those found deterministically (i.e., when time values are constant).

Most simulation studies of discrete manufacturing have been applied to the "push-method" wherein every station operates as fully as possible "pushing" the work out. This scheme is not generally used in modern factories, since large buffers between stations are usually required and there is no immediate incentive to keep from totally jamming up production or making unusable units, which often must then be discarded. The "pull-method," on the other hand, requires that the prior station (location $n - 1$) not "push" out a "product" until a particular station (location n) is ready to "pull" it in (there may be a buffer of 1, or very few, between the two stations). This scheme also allows any station to stop the workflow until a problem that occurs at that point gets fixed. The resulting increase in the quality of the final product is significant, sometimes incredibly so.

There will be at least one bottleneck station in every system, which generally controls the throughput. The type of system to be considered is asynchronous (i.e., one in which the stations have different cycle-times). Occasionally, one of the nonbottleneck stations takes a longer time than the bottleneck to get its work accomplished because of a variety of factors. Note that the uptime expected parameter used in the system design is intended to account for those variable conditions. Estimated behavior for a particular production batch utilizing random time performance is to be determined. This is accomplished by assuming that the system cycle-time for any production unit is equivalent to the longest time encountered at any workstation involved in the production of that unit. Thus, the expected output of a system will likely be somewhat less than the theoretical value determined using the methods described in Chap. 3.

Two commonly used statistical distributions will be applied, separately, to each workstation in the system. How the stations truly interact with each other is not known absolutely; however, a technique for investigating various possibilities will be discussed.

4.2 Applicable Discrete Event Distributions

Fabrication and assembly of electromechanical components, subassemblies, and assemblies is a discrete event occurrence (as opposed to, say, chemical processing, which is continuous). There are not as many discrete-event probability distribution types available as those that may be applied to

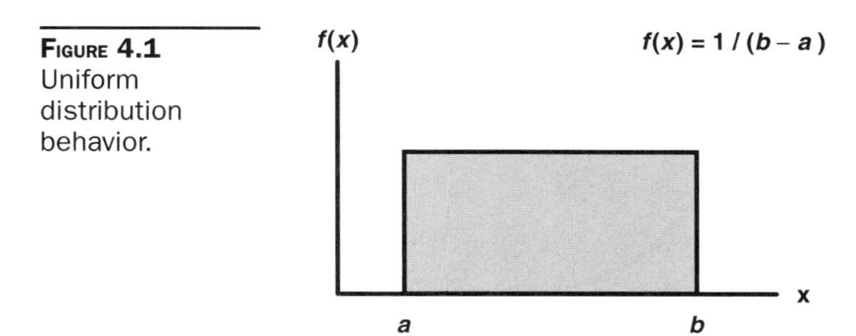

FIGURE 4.1 Uniform distribution behavior.

continuous events. In fact, there are three that can be readily applied to manufacturing systems:

The **uniform distribution**—every cycle-time between a minimum (a) and a maximum (b) has an equal likelihood of occurrence as shown in Fig. 4.1. The expected time is $(a + b)/2$. This is the worst-case scenario. It would be used only in a "rough order of magnitude" situation where the task times are very coarsely defined.

The **triangular distribution**—estimates for the minimum, maximum, and most-likely times are obtainable by some means. Fig. 4.2 exhibits the characteristics.

The **exponential distribution**—used in situations in which the random quantities do not depend upon any of their previous values as seen in Fig. 4.3.

The latter two distributions have been implemented in the system design procedure. Displaying random behavior begins with the stochastic evaluation data (see Fig. 4.4).

Both distribution types will be investigated. Similar characteristics will be seen; the exponential distribution generally

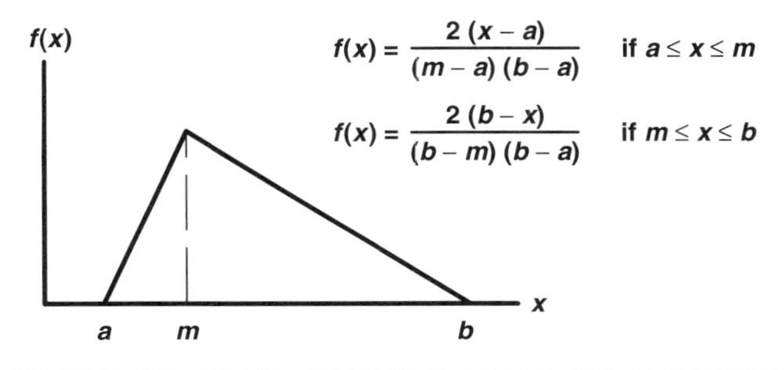

FIGURE 4.2 Triangular distribution behavior.

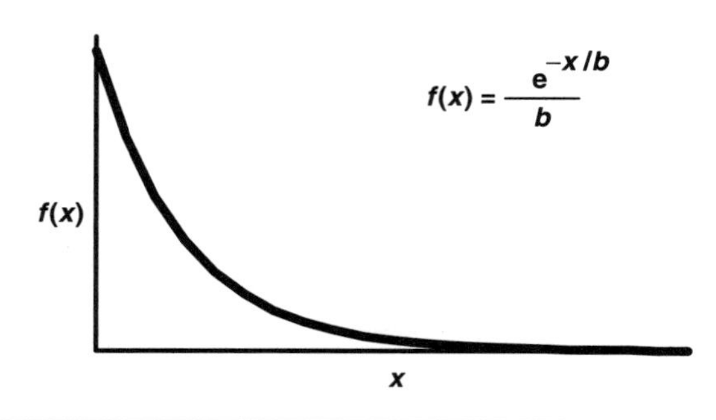

FIGURE 4.3 Exponential distribution behavior.

exhibits longer station time possibilities than the triangular distribution; therefore, the system throughput for that type will usually be less.

4.3 Using the Triangular Distribution

For this situation, a theoretical station time (designated m) is known and there is some insight into the values for the minimum a and maximum b station times. Recall that $f(x)$

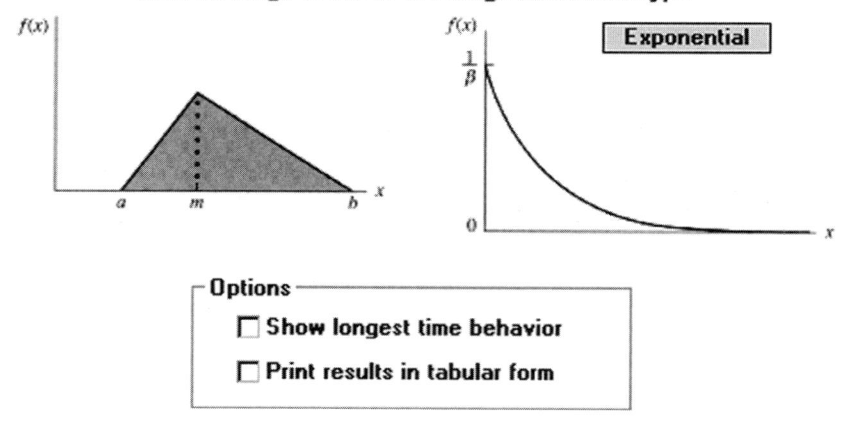

FIGURE 4.4 Options for stochastic analysis.

is known as the probability density function. One other extremely useful piece of information is also known: the expected value E for the station time. It is defined simply as the theoretical time divided by the (decimal) uptime expected ε for the station, or

$$E = m/\varepsilon$$

Recall that ε is a resource type parameter used in the system design procedure to establish deterministic workstation time characteristics.

For a triangular distribution, the expected value is

$$E = \frac{(a + m + b)}{3}$$

Some means to determine the minimum a and the maximum b needs to be defined. A ratio of *nominal to maximum time range* to *minimum to nominal time range* will be specified by

$$\beta = \frac{(b - m)}{(m - a)}$$

For production stations, β is likely to be quite large since processing times less than the theoretical (also called the nominal or most-likely) will seldom occur and would usually be only slightly lower in the worst case. For the stochastic evaluator portion of the system design procedure, $\beta = 200$ for all stations.

Since a schematic layout of a manufacturing system (see Fig. 3.17) exhibits the station expected time, E will be used as the basis for determining the minimum and maximum station times.

$$a = E\frac{(\beta + 2)\varepsilon - 3}{\beta - 1} = E\frac{202\,\varepsilon - 3}{199}$$

$$b = E\frac{3\beta - (2\beta + 1)\varepsilon}{\beta - 1} = E\frac{600 - 401\,\varepsilon}{199}$$

A plot of these relationships is exhibited in Fig. 4.5. Note that this graph is nondimensional and it can be applied to every

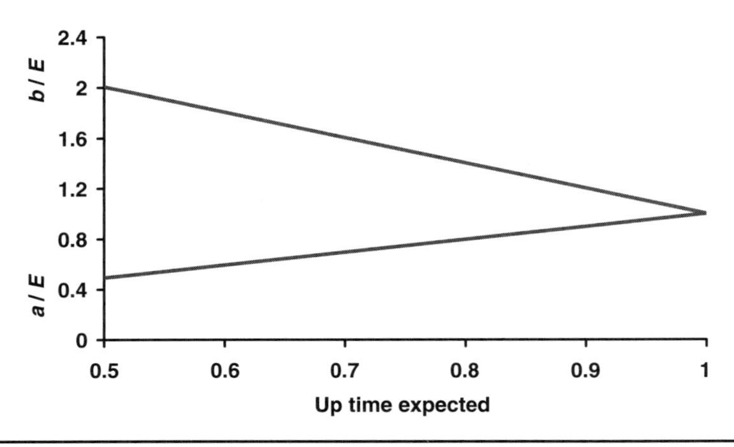

FIGURE 4.5 Range limits for prescribed uptime.

workstation by simply multiplying by the expected time E for that station. It is necessary to establish a and b since the "actual" station time t found during the random number analysis is defined as

$$t = a + (b - a)Q$$

To determine Q, the cumulative distribution function (CDF) for a triangular distribution is used. It will be calculated in two parts, each based upon the probability distribution function (PDF) defined earlier.

Range	CDF
$a \leq x \leq m$	$(\beta + 1)[(x - a)/(b - a)]^2$
$m \leq x \leq b$	$1 - (1 + 1/\beta)[(b - x)/(b - a)]^2$

Behavior when $\beta = 200$ is shown in Fig. 4.6. The random number generator program used in computers provides a stream of arbitrary numbers R, with values between 0 and 1, which is repeatable when the same random number seed (RNS) is used. Since the initial value of the CDF must be 0 and the final value must be 1, you can simply substitute R

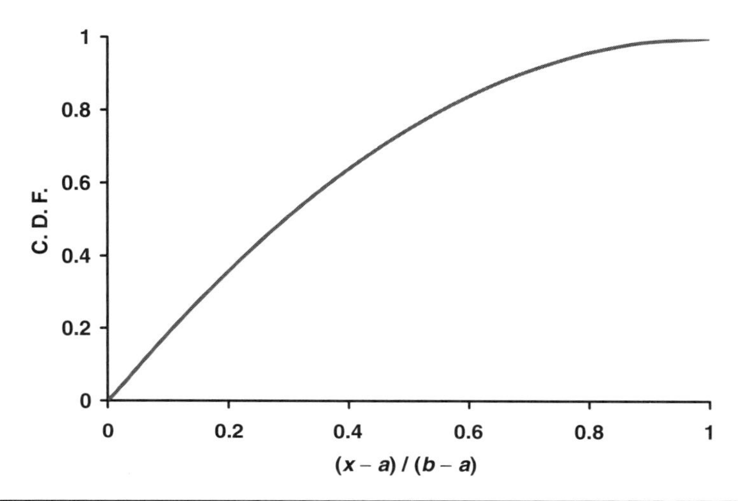

Figure 4.6 Cumulative distribution function for any point in a triangular distribution.

for CDF in the above equations and rearrange them to find the Q relationships:

Range	Q
$(0 \le R \le 1)/(\beta + 1)$	$Q = \sqrt{\dfrac{R}{\beta + 1}}$
$1/(\beta + 1) \le R \le 1$	$Q = 1 - \sqrt{\dfrac{\beta(1 - R)}{\beta + 1}}$

This information is essentially an "invert the axes plot" from that shown (Fig. 4.6) and is displayed in Fig. 4.7. This is the basis for a stochastic evaluation of a production, or other discrete event, system. The random number behavior of each station that can possibly contribute the longest station time allows us to evaluate the expected throughput of the system.

4.4 Application to a Manufacturing System

For any given manufacturing system (see Fig. 3.17), not necessarily all of the stations can contribute to the longest time;

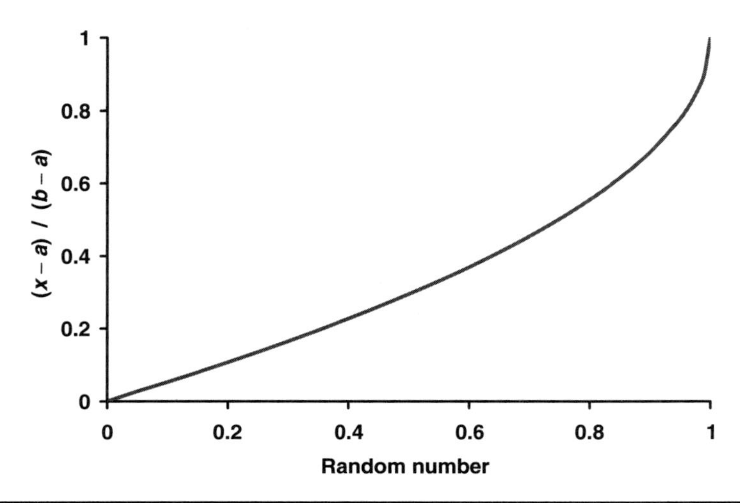

Figure 4.7 "Inverse" cumulative distribution function vs. position in range.

therefore, concentration will be placed only on those workstations that do affect system throughput.

A thorough investigation of the workstations determines that only 10 of them need further consideration. The top portion of Table 4.1 exhibits the various parameters (m, E, b) discussed above for those 10 stations; they are not all truly adjacent but will be treated as if they were in the following discussion. Note that two stations have the same "bottleneck" characteristics (shown in bold); for this example, they are replicated stations.

There are two options for the RNS at each station:

1. *Common value*—all stations use the same RNS. This is a widespread practice in simulations since it makes the computations considerably simpler. For present purposes, this scheme provides the "best-case" random number behavior.
2. *Unique value*—each station uses a different RNS, even for replicated stations. This arrangement provides the "worst-case" random number behavior.

What is sought is an estimate of the system throughput. The lower two-thirds of Table 4.1 exhibit the stochastic behavior for five different conditions using unique RNSs. Note how similar all of the average time and expected time values for each station turn out to be. In the general procedure, it is necessary to specify how many prior and/or subsequent stations influence the cycle-time of any station. In Table 4.1, the largest such number (9) for 10 stations has been selected. If 0 stations affecting cycle-time had been specified, the expected time would be exactly the same as the average time for each station. The minimum effect (for +/−1 station) is shown in Table 4.2. Note that the longest expected time occurs at the "second" station (MNL-3) under these conditions; now that's an unexpected result! Also, the projected throughput is 18.9% higher than the predicted worst-case capacity.

It normally takes the influence of approximately half the contributing stations to provide the maximum effect (i.e., establish the lowest throughput from the system). For the particular system being investigated, the influence of plus or minus four or five stations provides the worst case. However, to readily establish the worst case, it is wise to always use one less than the total number of longest station time contributing stations (e.g., use a value of 5 for a 6 potentially longest-time station system).

The yearly production volume capacity of the system can be calculated as

$$V = \frac{3600(8) S_D D_Y}{T_{CYC}}$$

where 3600 = seconds per hour
8 = hours per shift
S_D = shifts per day
D_Y = work days per year
T_{CYC} = cycle-time (seconds per unit)

Four graphs exhibiting the stochastic behavior of a manufacturing system are available. For the system being considered,

1. average workstation time (Fig. 4.8) shows how early pertubations in the average time settle out by about 1000 units and also that the worst-case system time is larger than the longest station time.

Resource data set: **ACMRES-B**
Required yearly production is 425000
Maximum stations affecting cycle-time +/– 9

Station	MNL-2	MNL-3	MNL-4	P55-1	MNL-6	MNL-6	MNL-7	MNL-9	MNL-9	FXD-7
Fst Tsk	5	6	7	8	10	10	15	**17**	**17**	21
Lst Tsk	5	6	7	9	11	11	16	**19**	**19**	21
NomTime	23.00	23.00	23.00	24.50	21.25	21.25	19.50	**24.00**	**24.00**	24.00
CycTime	26.03	26.03	26.03	26.02	24.05	24.05	22.07	**27.16**	**27.16**	25.49
MaxTime	32.13	32.13	32.13	29.08	29.68	29.68	27.24	**33.52**	**33.52**	28.49

27.23 best-case RN system cycle-time
30.37 worst-case RN system cycle-time

RNS	6	12	18	24	30	36	42	**48**	**54**	60
avg tme	26.07	26.02	25.90	26.11	24.22	24.13	22.20	**27.31**	**27.18**	25.52
exp. Tme	30.32	30.32	30.32	30.32	30.32	30.32	30.32	**30.32**	**30.32**	30.32
RNS	58	56	54	52	50	48	46	**44**	**42**	40
avg tme	26.12	25.94	26.09	26.02	24.10	24.00	22.10	**27.31**	**27.11**	25.44
exp. Tme	30.36	30.36	30.36	30.36	30.36	30.36	30.36	**30.36**	**30.36**	30.36
RNS	39	38	37	36	35	34	33	**32**	**31**	30
avg tme	26.08	25.84	26.00	25.97	23.88	24.18	22.04	**27.21**	**27.11**	25.50

Station	MNL-2	MNL-3	MNL-4	P55-1	MNL-6	MNL-6	MNL-7	MNL-9	MNL-9	FXD-7
exp. Tme	30.16	30.16	30.16	30.16	30.16	30.16	30.16	**30.16**	**30.16**	30.16
RNS	28	26	24	22	20	18	16	**14**	**12**	10
avg tme	25.97	26.12	26.06	26.01	23.98	24.04	22.05	**27.15**	**27.22**	25.52
exp. Tme	30.24	30.24	30.24	30.24	30.24	30.24	30.24	**30.24**	**30.24**	30.24
RNS	5	10	15	20	26	30	35	**40**	**45**	50
avg tme	26.16	26.13	26.07	26.01	24.04	23.93	22.21	**27.17**	**27.13**	25.52
exp. Tme	30.32	30.32	30.32	30.32	30.32	30.32	30.32	**30.32**	**30.32**	30.32

Summary

Station	MNL-2	MNL-3	MNL-4	P55-1	MNL-6	MNL-6	MNL-7	MNL-9	MNL-9	FXD-7
Uptime	88.4%	88.4%	88.4%	94.2%	88.4%	88.4%	88.4%	**88.4%**	**88.4%**	94.2%
avg tme	26.08	26.01	26.02	26.03	24.04	24.06	22.12	**27.23**	**27.15**	25.50
exp. tme	30.28	30.28	30.28	30.28	30.28	30.28	30.28	**30.28**	**30.28**	30.28

Yearly capacity for ave tme is 507666
Yearly capacity for exp. tme is 456541

TABLE 4.1 Stochastic Results (Maximum Influence) Using Triangular Distributions

Resource data set: **ACMRES-B**
Required yearly production is 425000
Maximum stations affecting cycle-time +/– 1

RNS	98	96	94	92	90	88	86	**84**	**82**	80
avg tme	26.19	25.99	26.02	25.98	24.04	24.28	22.13	**27.04**	**27.02**	25.46
exp. tme	27.66	28.62	28.15	27.37	26.61	24.80	26.08	**27.25**	**27.54**	26.49
RNS	68	67	66	65	64	63	62	**61**	**60**	59
avg tme	26.03	26.10	25.98	26.02	24.01	24.14	21.91	**27.19**	**27.48**	25.58
exp. tme	27.60	28.57	28.24	27.41	26.59	24.62	26.03	**27.49**	**27.85**	26.82
RNS	357	368	379	567	578	589	600	**611**	**702**	7665
avg tme	26.00	26.00	26.15	26.05	24.11	24.22	22.02	**27.31**	**26.98**	25.54
exp. tme	27.54	28.62	28.25	27.59	26.72	24.79	26.15	**27.35**	**27.66**	26.53
RNS	4	8	12	16	20	24	28	**32**	**36**	40
avg tme	25.96	25.94	25.91	26.02	23.79	23.96	22.08	**27.13**	**27.05**	25.53
exp. tme	27.44	28.40	28.06	27.34	26.49	24.52	25.97	**27.30**	**27.59**	26.55
RNS	5	15	25	35	45	55	65	**75**	**85**	95
avg tme	25.99	26.02	26.10	26.02	24.07	24.15	22.08	**27.24**	**26.97**	25.54
exp. tme	27.54	28.61	28.26	27.52	26.61	24.71	26.05	**27.27**	**27.58**	26.49

Summary

Station	**MNL-2**	**MNL-3**	**MNL-4**	**P55-1**	**MNL-6**	**MNL-6**	**MNL-7**	**MNL-9**	**MNL-9**	**FXD-7**
uptime	88.4%	88.4%	88.4%	94.2%	88.4%	88.4%	88.4%	**88.4%**	**88.4%**	94.2%
avg tme	26.03	26.01	26.03	26.02	24.00	24.15	22.05	**27.18**	**27.10**	25.53
exp. tme	27.56	28.56	28.19	27.44	26.61	24.69	26.06	**27.33**	**27.64**	26.58

Yearly capacity for ave tme is 508601
Yearly capacity for exp. tme is 483969

Table 4.2 Stochastic Results (Minimum Influence) Using Triangular Distributions

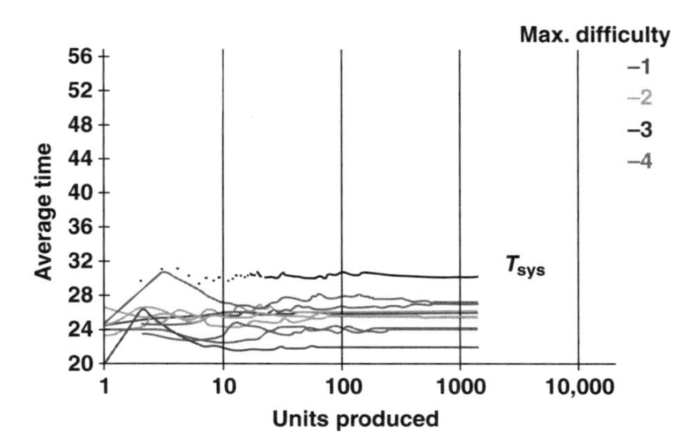

Figure 4.8 Average workstation throughput time—triangular distributions.

2. stochastic station time (Fig. 4.9) shows how the time for each station varies randomly, but always within its range, and that one of the stations will have the longest time.
3. ratio of total stochastic station time to total expected station time (Fig. 4.10) shows that the actual time required for production is greater than or equal to the nominal.
4. ratio of actual production to required production (Fig. 4.11) exhibits the possibility that either the random number worst case or the random number best case might be less than the requirement.

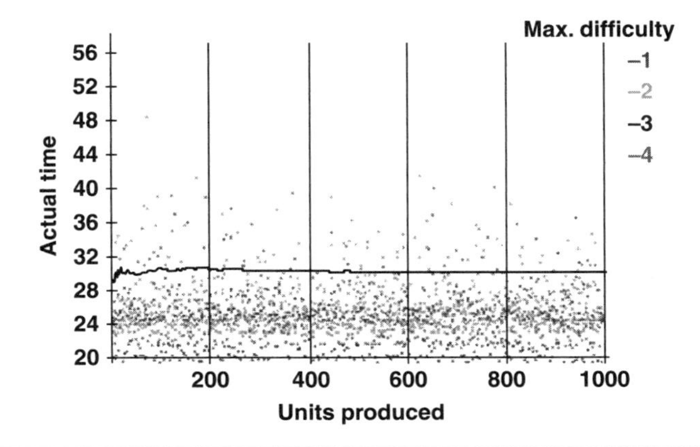

Figure 4.9 Stochastic station times—triangular distributions.

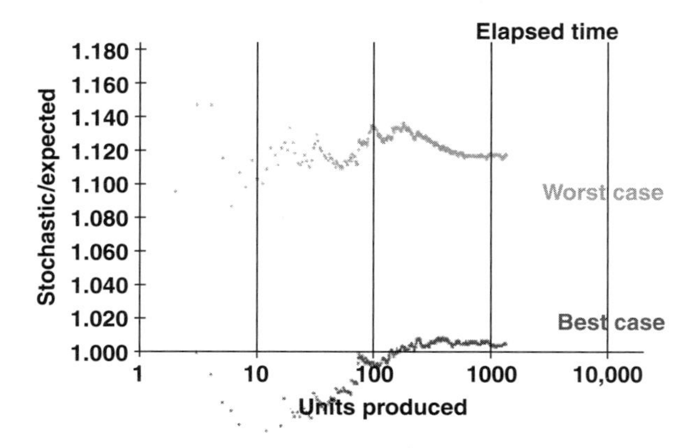

FIGURE 4.10 Stochastic time ratios—triangular distributions.

4.5 Using the Exponential Distribution

Conditions very similar to those already discussed for the triangular distribution case will be found for an application of the exponential distribution to the behavior of a production system. For this situation, a theoretical station time (designated m) is known and there is some insight into the values for the minimum a and maximum b station times. Again, note that $f(x)$ is the probability density function. Also recognize one other extremely useful piece of information: the

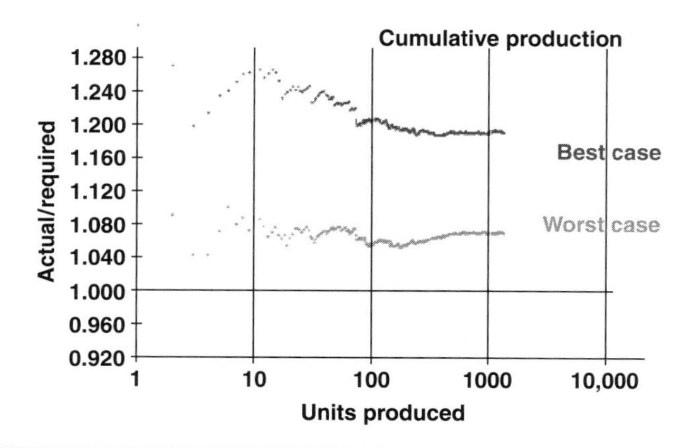

FIGURE 4.11 Stochastic production ratios—triangular distributions.

expected value E for the station time. It is defined simply as the theoretical time divided by the (decimal) uptime expected ε for the station, or

$$E = m/\varepsilon$$

Recall that ε is a resource type parameter used in the system design procedure to establish deterministic workstation time characteristics.

For an exponential distribution, the expected value is

$$E = \beta$$

Note that this β is not the same as that defined for the triangular distribution; it is defined to be a fundamental parameter for an exponential distribution. The actual value is specified as

$$\beta = E(1 - \varepsilon)$$

Since a schematic layout of a manufacturing system (see Fig. 3.17) exhibits the station expected time, you can use that E as the basis for determining the minimum and maximum station times. Here, the same minimum station time (which is slightly less than the nominal time) as for the triangular distribution will be used:

$$a = E\frac{202\,\varepsilon - 3}{199}$$

For practical purposes, there is such a slight difference between a and m (the nominal time value) that the latter (m) will be utilized as the minimum time value.

The maximum time b is actually mathematically infinite (i.e., seriously large), but a very good approximation will be

$$b = E \times \varepsilon\left[1 - \left(\frac{1}{\varepsilon} - 1\right)\log(0.0001)\right]$$

The minimum and maximum station times can be plotted as shown in Fig. 4.12. Note that this graph is nondimensional and it can be applied to every workstation by simply multiplying by the expected time E for that station. It is necessary

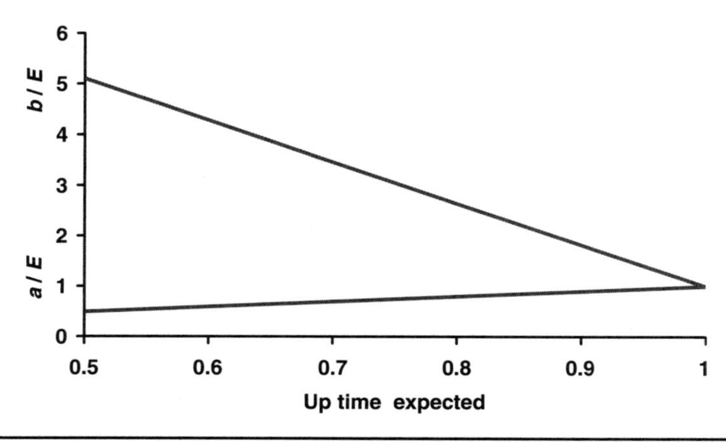

FIGURE 4.12 Range limits for prescribed uptime.

to establish a and b since the "actual" station time t found during the random number analysis is

$$t = a + (b - a)\,Q$$

The exponential distribution has the equation

$$f\,(x) = \frac{1}{\beta} e^{-x/\beta}$$

Fig. 4.13 is a plot of this function for two typical uptime expectations. To determine Q, the CDF for an exponential distribution is used:

$$\text{CDF} = 1 - e^{-x/\beta}$$

The behavior of this equation is exhibited in Fig. 4.14. The random number generator program used in computers provides a stream of arbitrary numbers R, with values between 0 and 1, which is repeatable when the same RNS is used. Since the initial value of the CDF must be 0 and the final value must be 1, R will simply be substituted for CDF in the above

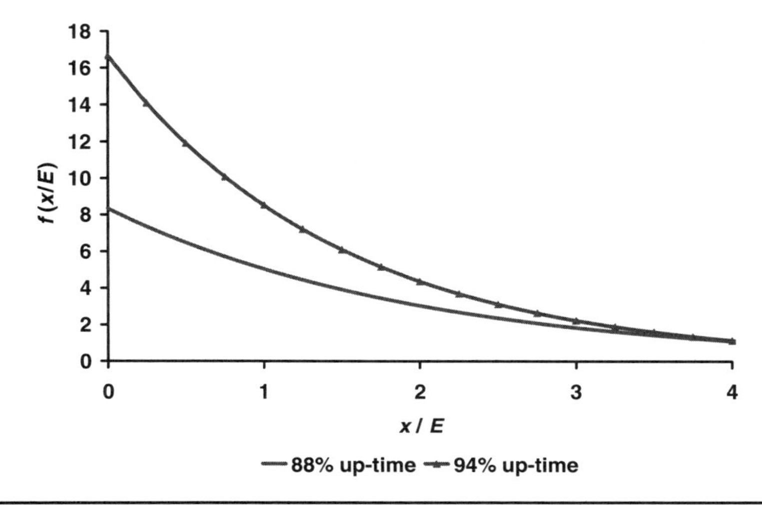

FIGURE 4.13 Application of exponential distribution.

equations, which are rearranged to find the Q relationship:

$$Q = \frac{x}{E} = -\ln(1 - R)$$

This information is essentially an "invert the axes plot" from Fig. 4.14 as seen in Fig. 4.15. This is the basis for a

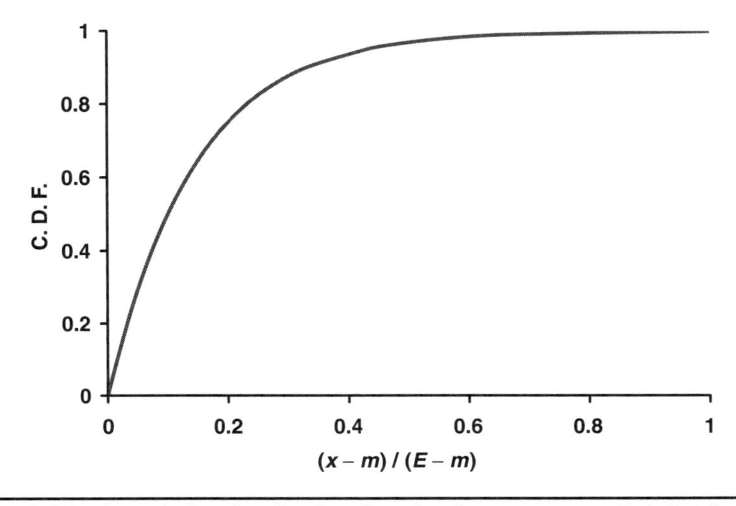

FIGURE 4.14 Cumulative distribution function for an exponential distribution.

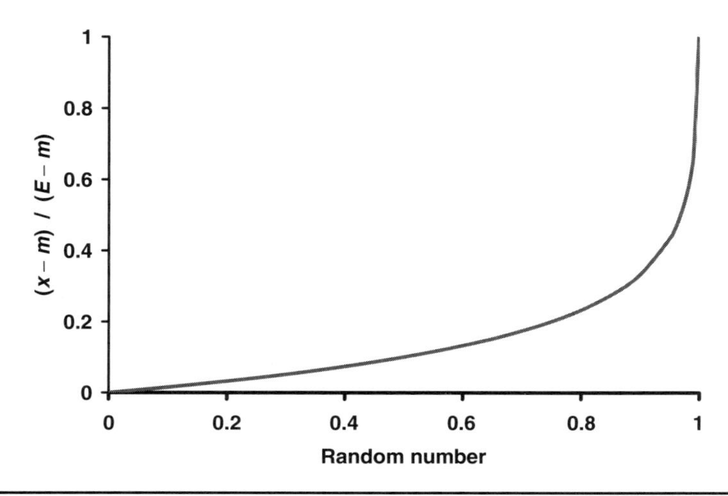

FIGURE 4.15 "Inverse" cumulative distribution function for an exponential distribution.

stochastic evaluation of a manufacturing, or other discrete event, system. The random number behavior of each station that can possibly contribute the longest station time permits evaluation of the throughput of the system. Table 4.3 exhibits the worst-case condition. When only the nearest contributing workstation affects the throughput, the expected behavior is shown in Table 4.4. Expected production is 6.2% higher than when the maximum other station influence occurs (see Table 4.3).

Four graphs exhibiting the stochastic behavior of a manufacturing system are available. Examples for the system being considered are as follows:

1. Average workstation time (Fig. 4.16) shows how early pertubations in the average time settle out by about 1000 units and also that the worst-case system time is larger than the longest station time
2. Stochastic station time (Fig. 4.17) shows how the time for each station varies randomly, but always within its range, and that one of the stations will have the longest time.
3. Ratio of total stochastic time to total expected time (Fig. 4.18) shows that the actual station time required for production is greater than or equal to the nominal station time.

Resource data set : **ACMRES-B**
Required yearly production is 425000
Maximum stations affecting cycle-time +/− 10

Station	MNL-1	MNL-2	MNL-3	MNL-4	P55-1	MNL-6	MNL-6	MNL-7	MNL-9	MNL-9	FXD-7
Fst Tsk	4	5	6	7	8	10	10	15	**17**	**17**	21
Lst Tsk	4	5	6	7	9	11	11	16	**19**	**19**	21
NomTime	17.50	23.00	23.00	23.00	24.50	21.25	21.25	19.50	**24.00**	**24.00**	24.00
CycTime	19.80	26.03	26.03	26.03	26.02	24.05	24.05	22.07	**27.16**	**27.16**	25.49
MaxTime	38.71	50.88	50.88	50.88	38.50	47.01	47.01	43.14	**53.09**	**53.09**	37.71
27.23 best-case RN system cycle-time 30.40 worst-case RN system cycle-time											
RNS	101	103	105	201	203	205	301	303	**305**	**401**	403
avg tme	19.66	26.13	25.74	25.97	26.09	23.86	24.06	22.30	**26.96**	**27.43**	25.48
exp. tme	30.14	30.14	30.14	30.14	30.14	30.14	30.14	30.14	**30.14**	**30 14**	30.14
RNS	502	504	506	602	604	606	702	704	**706**	**602**	804
avg tme	19.76	26.19	26.03	25.94	25.96	24.14	23.94	21.96	**27.11**	**27.22**	25.44
exp. tme	30.33	30.33	30.33	30.33	30.33	30.33	30.33	30.33	**30.33**	**30.33**	30.33
RNS	96	94	92	90	88	86	84	82	**80**	**76**	76

Table 4.3 Stochastic Results (Maximum Influence) Using Exponential Distributions *(Continued)*

Station	MNL-1	MNL-2	MNL-3	MNL-4	P55-1	MNL-6	MNL-6	MNL-7	MNL-9	MNL-9	FXD-7
avg tme	19.90	25.90	26.07	25.79	26.04	24.06	23.91	21.98	**27.36**	**27.20**	25.55
exp. tme	30.17	30.17	30.17	30.17	30.17	30.17	30.17	30.17	**30.17**	**30.17**	30.17
RNS	74	72	70	68	66	64	62	60	**58**	**56**	54
avg tme	19.74	25.96	26.34	25.87	26.13	24.08	24.04	22.16	**27 16**	**27.24**	25.45
exp. tme	30.51	30.51	30.51	30.51	30.51	30.51	30.51	30.51	**30.51**	**30.51**	30.51
RNS	25	50	75	100	125	150	175	200	**225**	**250**	275
avg tme	19.63	26.10	26.20	26.21	26.02	23.99	24.10	21.89	**27.20**	**27.26**	25.53
exp. tme	30.37	30.37	30.37	30.37	30.37	30.37	30.37	30.37	**30.37**	**30.37**	30.37

Summary

station	MNL-1	MNL-2	MNL-3	MNL-4	P55-I	MNL-6	MNL-6	MNL-7	MNL-9	MNL-9	FXD-7
uptime	88.4%	88.4%	88.4%	88.4%	94.2%	88.4%	88.4%	88.4%	**86.4%**	**66.4%**	94.2%
avg tme	19.74	26.06	26.08	25.96	26.05	24.03	24.01	22.06	**27.16**	**27.27**	25.49
exp. tme	30.30	30.30	30.30	30.30	30.30	30.30	30.30	30.30	**30.30**	**30 30**	30.30

Yearly capacity for ave tme is 506957
Yearly capacity for exp. tme is 456165

Table 4.3 Stochastic Results (Maximum Influence) Using Exponential Distributions *(Continued)*

Resource data set: **ACMRES-B**
Required yearly production is 425000
Maximum stations affecting cycle-time +/– 1

RNS	7	12	17	22	27	32	37	42	**47**	**52**	57
avg tme	19.76	25.92	25.98	26.07	26.05	23.90	24.72	22.10	**27.01**	**27.03**	25.57
exp. tme	26.06	27.55	28.48	28.22	27.51	26.77	24.91	26.23	**27.23**	**27.57**	26.58
RNS	98	97	96	95	94	93	92	91	**90**	**89**	88
avg tme	19.79	26.06	26.15	26.07	26.04	24.02	24.09	22.15	**27.20**	**26.80**	25.64
exp. tme	26.15	27.72	28.76	28.30	27.47	26.53	24.75	26.09	**27.25**	**27.57**	26.48
RNS	86	84	82	80	78	76	74	72	**70**	**68**	66
avg tme	19.81	26.06	26.01	26.01	25.91	24.05	23.98	21.89	**27.03**	**26.94**	25.53
exp. tme	26.15	27.55	28.52	28.12	27.38	26.43	24.55	25.85	**27.18**	**27.48**	26.49
RNS	63	60	57	54	51	48	45	42	**39**	**36**	33
avg tme	19.87	25.82	26.02	26.10	26.03	24.09	23.95	22.04	**27.23**	**27.03**	25.46
exp. tme	25.92	27.45	28.47	28.24	27.59	26.67	24.64	26.00	**27.32**	**27.60**	26.50
RNS	2	8	14	20	26	32	38	44	**50**	**56**	62
avg tme	19.89	25.96	25.81	25.89	26.05	23.80	23.93	22.15	**27.35**	**26.84**	25.52
exp. tme	26.05	27.48	28.43	28.05	27.28	26.52	24.53	26.07	**27.30**	**27.65**	26.45

Summary

Station	MNL-1	MNL-2	MNL-3	MNL-4	P55-1	MNL-6	MNL-6	MNL-7	MNL-9	MNL-9	FXD-7
uptime	88.4%	88.4%	88.4%	88.4%	94.2%	88.4%	88.4%	88.4%	**88.4%**	**88.4%**	94.2%
avg tme	19.83	25.96	25.99	26.03	26.02	23.97	24.13	22.07	**27.16**	**26.93**	25.54
exp. tme	26.07	27.55	28.53	28.18	27.45	26.58	24.68	26.05	**27.26**	**27.57**	26.50

Yearly capacity for ave tme is 508891
Yearly capacity for exp. tme is 484509

Table 4.4 Stochastic Results (Minimum Influence) Using Exponential Distributions

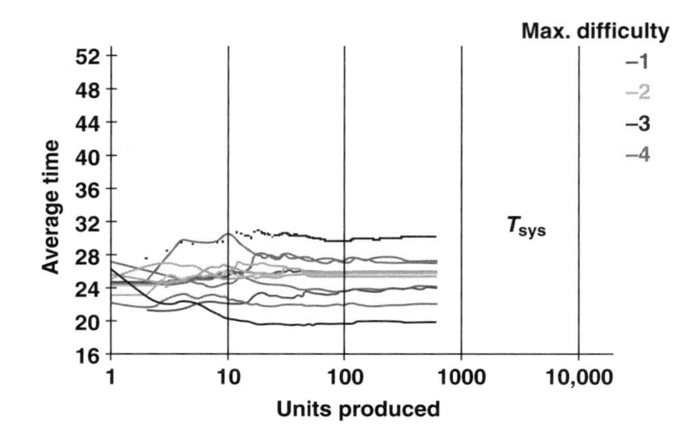

FIGURE 4.16 Average workstation throughput time—exponential distributions.

4. Ratio of actual production to required production (Fig. 4.19) shows that either the random number worst case or the random number best case might not equal the requirement.

Since the availability factor (portion of the total time available) for the synthesized system was 0.887 (see schematic in Fig. 3.17), the stochastic behavior of both distributions will be quite similar. Such will not be the case

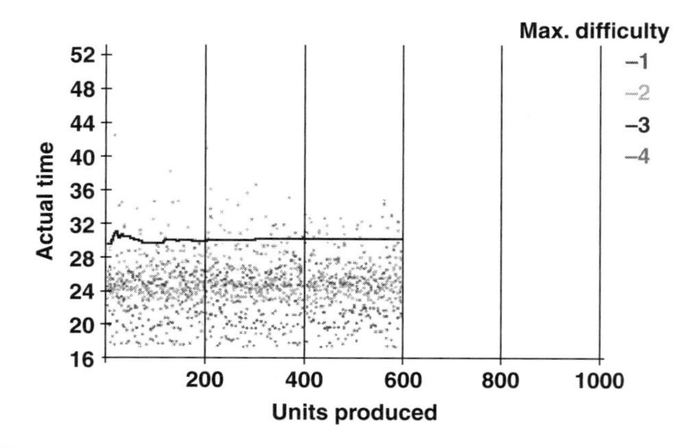

FIGURE 4.17 Stochastic station times—exponential distributions.

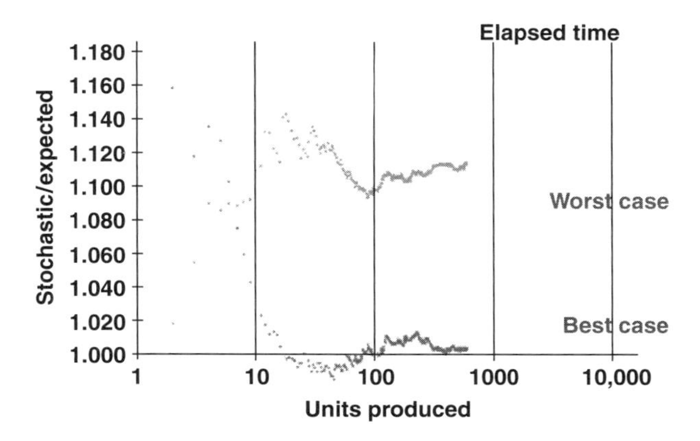

FIGURE 4.18 Stochastic time ratios—exponential distributions.

when operating near the maximum time for a shift. Fig. 4.20 exhibits the portion of the total events count for the longest expected time that occurred at each station.

A fundamental question can now be asked. "Is it possible to determine the condition for which the expected worst-case production still meets the requirements?" Fig. 4.21 (for the particular application being studied) shows the maximum allowable availability factor for a system such that production will be satisfactory.

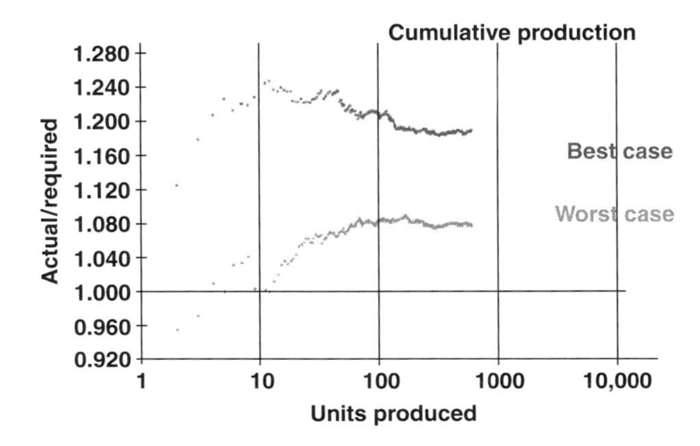

FIGURE 4.19 Stochastic production ratios—exponential distributions.

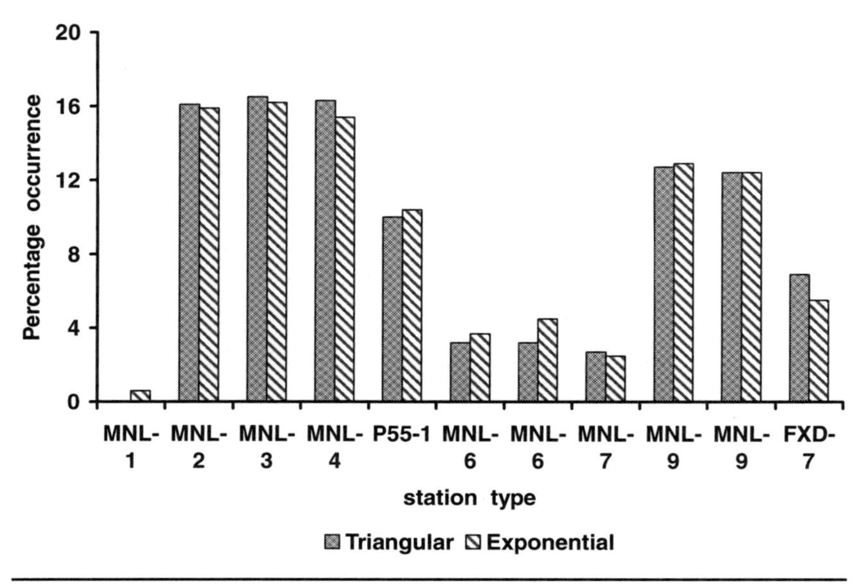

Figure 4.20 Count of longest station times.

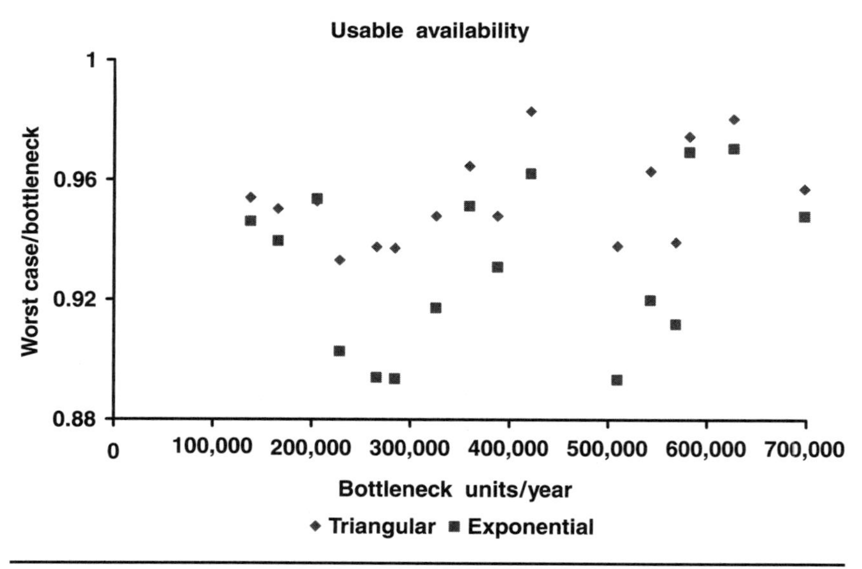

Figure 4.21 Favorable production requirements.

4.6 Application to Synthesis of Systems

It is now reasonable to ask whether the information just described can be made available at system design time. Since the primary interest is to determine the best estimate of system production volume, establishing the stochastic "worst-case" and "best-case" conditions will be necessary. There will be occasions when the stochastic production does not satisfy the required quantity; such conditions normally appear at or near the longest time for each shift available.*

For the **best-case** random number behavior (the *same* RNS is used for *all* stations), one of two cases is occurring:

1. All stations have been assigned the same resource type—system cycle-time equals the bottleneck expected time.

   Or

2. At least one station is a different resource type *and* its minimum time is greater than the bottleneck station's minimum time as shown in Fig. 4.22 (only the two station types with the largest expected time E need to be evaluated).

Since the **same RNS** is used for all stations, identical values for the random number will be generated for all stations. Fig. 4.22 shows that at times the bottleneck station (**MNL-1**) will not have the longest station time. The system cycle-time for this condition (for triangular distributions) is calculated using

$$T_{SYS} = E_{NBN} + \left[\frac{\beta + 1}{3\beta}\right] \frac{[b_{NBN} - b_{BN}]^3}{[b_{NBN} - a_{NBN} - b_{BN} + a_{BN}]^2}$$

where BN = the bottleneck station
NBN = the nonbottleneck station

* The example used in this discussion is not the same as that given earlier in this chapter; it was selected because it emphasizes the conditions being evaluated.

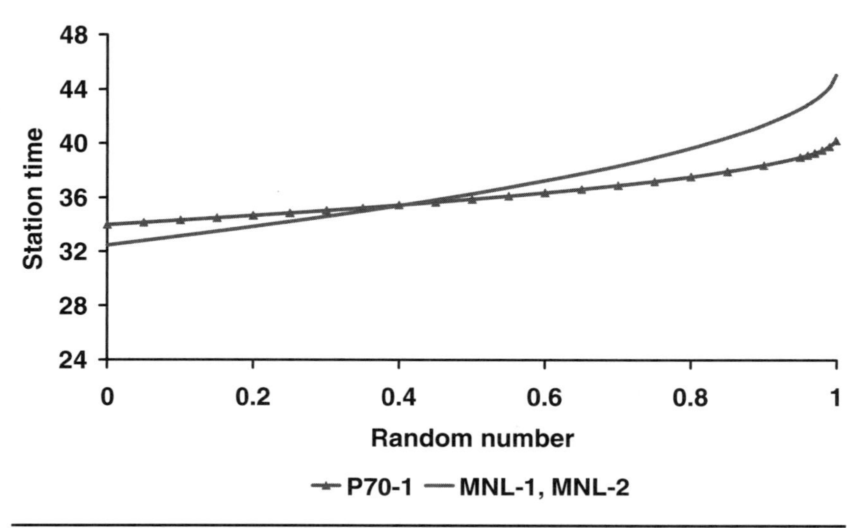

FIGURE 4.22 Same random number seed behavior.

When exponential distributions are used, the necessary condition (where LG denotes the longest time station) is

$$T_{\mathrm{SYS}} = E_{\mathrm{LG}} + [(E_{\mathrm{BN}} - m_{\mathrm{BN}}) - (E_{\mathrm{LG}} - m_{\mathrm{LG}})]e^{\left[\frac{m_{\mathrm{LG}} - m_{\mathrm{BN}}}{(E_{\mathrm{LG}} - m_{\mathrm{LG}}) - (E_{\mathrm{BN}} - m_{\mathrm{BN}})}\right]}$$

For the **worst-case** random number behavior (a *different* RNS is used for *each* station), considerably more intermediate information must be determined. An example of the conditions encountered is shown in Fig. 4.23.

To date, no mathematically provable equation describing this situation has been found. However, the following has been shown to be an excellent heuristic (differences between calculated and stochastic times are usually less than 0.5%; see Table 4.1) for the triangular distribution case:

$$T_{\mathrm{SYS}} \cong E_{\mathrm{BN}}\left\{1 + \frac{7}{9}\sqrt{\frac{(1 - \varepsilon_{\mathrm{BN}})(\beta + 1)}{3\beta\, E_{\mathrm{BN}}}\left[\sum_{K} \frac{(b_k - E_{\mathrm{BN}})^3}{(b_k - a_k)^2} - \frac{(b_{\mathrm{BN}} - E_{\mathrm{BN}})^3}{(b_{\mathrm{BN}} - a_{\mathrm{BN}})^2}\right]}\right\}$$

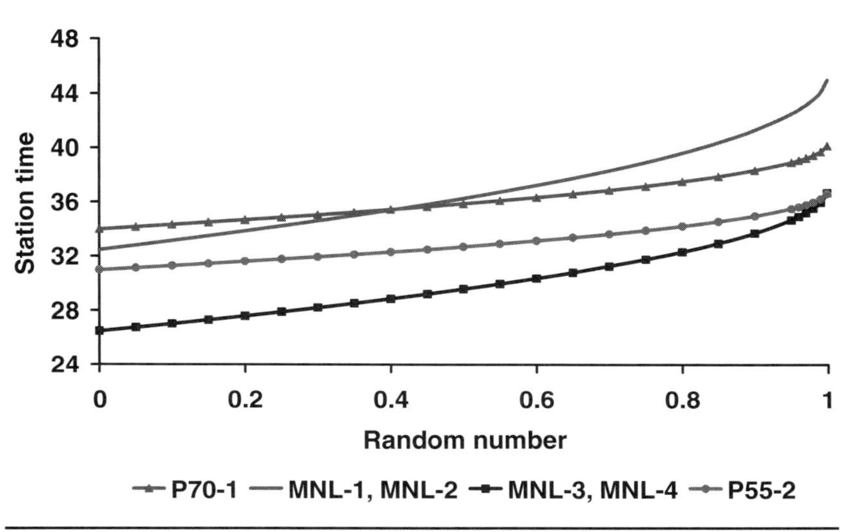

Figure 4.23 Unique random number seeds behavior.

When the exponential distribution case is desired, the relationship becomes

$$T_{SYS} = E_{BN} + \sum_{i=1}^{N} (E_i - m_i)\, e^{-\left(\frac{E_{BN} - m_i}{E_i - m_i}\right)} \qquad i \neq \text{BN, only}$$

The characteristics for triangular and exponential distributions have been added to the advanced system design procedure and can be seen near the top in Fig. 3.17 from which you will note the following:

Production requirement	425000/year
Bottleneck system capacity	509011/year (+19.8%)
RN worst-system capacity, triangular	477417/year (+12.3%)
RN best-system capacity, triangular	508053/year (+19.5%)
RN worst-system capacity, exponential	454700/year (+6.9%)
RN best-system capacity, exponential	507719/year (+19.4%)

There will be occasions when the worst-case system capacity will be less than the production requirement. Depending upon the magnitude of the discrepancy, some system parameter(s) may need to be changed.

4.7 Summary

A method for analyzing the random behavior of production systems has been described. Comparisons with steady-state (or deterministic) performance have been made. The ability to estimate stochastic throughput has become a useful feature of the advanced system design procedure.

CHAPTER 5

Initial Look at System Configurations

5.1 Introduction

While the schematic layout (which can represent a linear system) exhibited thus far contains much useful data, many people want to see at least a "stick-model" representation of what such a system would look like. While complementary software techniques allow very realistic pictures of systems to be created, here the intent is to provide only an initial look at the geometric relationships of the stations in a particular system.

5.2 Geometric Layouts

There are four layout formats generally used for production systems. The names come from the way they look from a "bird's eye view":

1. **Linear**—all stations are in line with action proceeding from one end to the other.

2. **Closed loop (without spacing)**—action proceeds clockwise or counterclockwise.
3. **Closed loop (with spacing)**—action proceeds clockwise or counterclockwise.
4. **"U" cell**—action proceeds clockwise or counterclockwise.

Each type possesses particular advantages depending upon many factors (e.g., space available, number of components, "comfortableness" factor). As stated above, the pictures to be created here will use only "stick" models; such outlines will provide the essential information for an initial geometric look at a system.

5.3 Schematic Layout Basis

Use Fig. 3.17 as the basic schematic for the system to be exhibited. Two important characteristics displayed in this schematic should be emphasized in the geometric layout:

1. Station(s) with the longest time (the "bottleneck")—defined by the maximum time shown in the system—will be denoted by a solid circle.
2. Station(s) with the maximum required effort—defined by the maximum effort for a system workstation, which usually defines where to put the "best" of the prescribed resource-type—will be denoted by a pale circle with a dark outline.

The outline box dimensions (width, depth, and height) for each of the resource types that appear in a system need to be specified in a unique data file. Then, the "rough cut" picture of the system geometry can be created. In all cases, the pictures include

manual stations—shown with solid fill in overhead views.

fixed automation stations—shown with lower left to upper right diagonal lines in overhead views.

programmable automation stations (robots)—shown with upper left to lower right diagonal lines in overhead views.

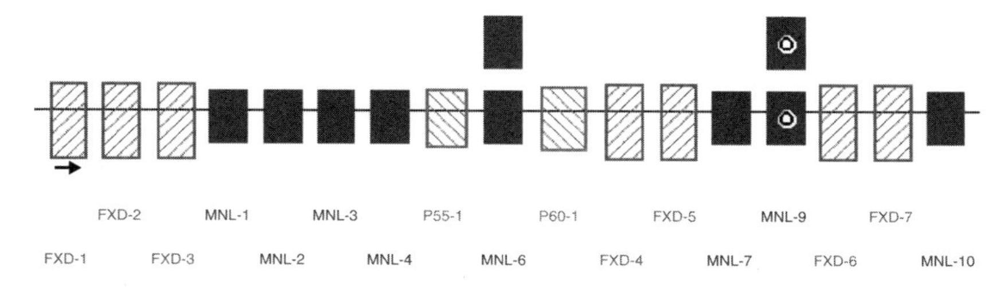

Figure 5.1 Linear layout for example system.

For convenience, counterclockwise work movement is assumed. The start station is identified by the arrow.

5.4 Linear System Layout

In this case, the system will be shown as a straight line with the work flowing from left to right or right to left (which are mirror images); see Fig. 5.1, for example. The bottleneck station also requires the maximum effort for this example, but this will not always be the case. For this system (not in the general case), two people working in parallel determine the "worst-case" condition; they require the maximum station time and maximum effort.

5.5 Closed Loop System—Without Spacing

Depending upon the product and its components, it may be possible to have a tightly packed system. For loop systems, there can be choices for the (x to y) shape factor such as the following:

	Number of stations				Dimension ratio (x/y)
	Sec. 1	Sec. 2	Sec. 3	Sec. 4	
Shape 1	4	6	3	7	0.53
Shape 2	4	6	4	6	0.65
Shape 3	5	5	4	6	0.79
Shape 4	5	5	5	5	0.96
Shape 5	6	4	5	5	1.16
Shape 6	6	4	6	4	1.42
Shape 7	7	3	6	4	1.73

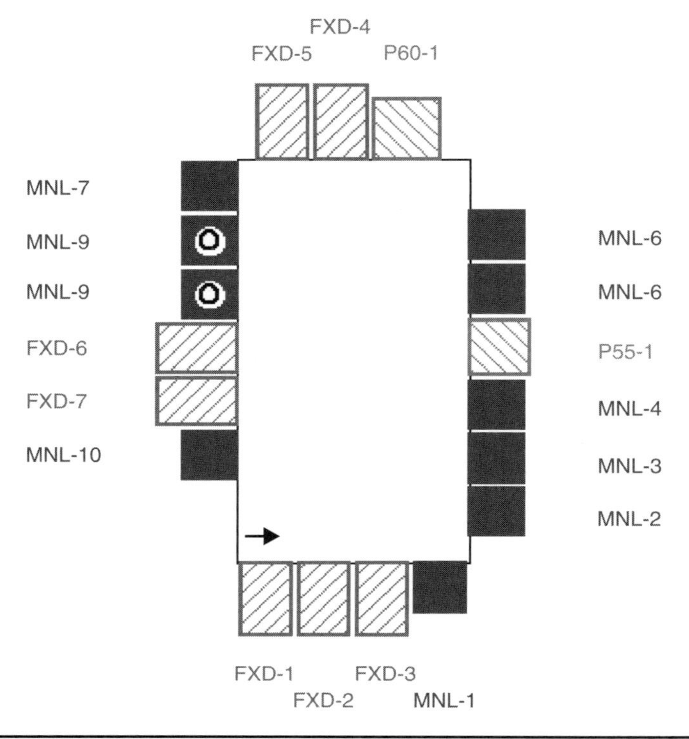

Figure 5.2 Tight loop system #1.

This is an unusually large number of choices for such a "small" system, but the variations do provide some interesting possibilities. For the minimum size ratio (0.53), Fig. 5.2 ensues. If a nearly square system (0.96) is desired, it would look like the one shown in Fig. 5.3. If the largest shape factor (1.73) is selected, Fig. 5.4 exhibits how the system would appear. These "tightly packed" systems, shown in rectangular mode, could also easily be circular if the number of workstations is (usually) 10 or less. While the flow of work is shown going counterclockwise, it could be clockwise in any of the systems shown.

5.6 Closed Loop System—With Spacing

When a system contains numerous workstations, and/or the components require significant space, the system will need to be spread out. A requirement for small in-process buffers

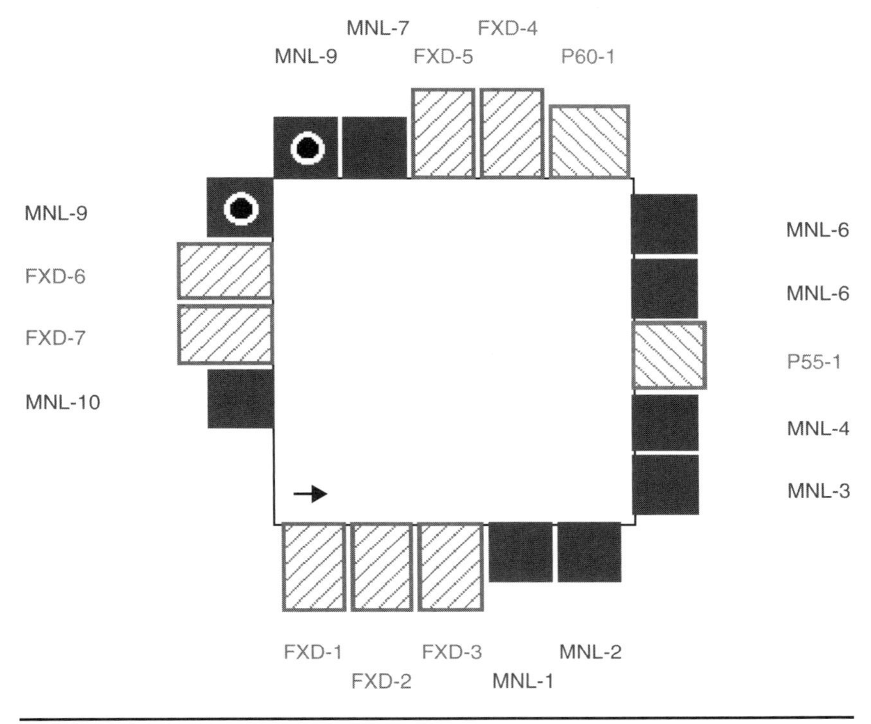

FIGURE 5.3 Tight loop system #2.

between the stations is also a possibility. For the example system of Fig. 3.17, the shape factor options are as follows:

	Number of stations				Dimension ratio (*x*/*y*)
	Sec. 1	Sec. 2	Sec. 3	Sec. 4	
Shape 1	4	5	3	7	0.59
Shape 2	4	5	4	6	0.73
Shape 3	5	4	4	6	0.91
Shape 4	5	4	5	5	1.12
Shape 5	6	3	5	5	1.39
Shape 6	6	3	6	4	1.73

These shape factors are nearly the same as those for the tightly packed case. As an example, select two different (intermediate ratio) cases. Fig. 5.5 shows the 0.73 size ratio case while the 1.39 size ratio condition is exhibited in Fig. 5.6. The spacing shown in these initial geometric layouts is based

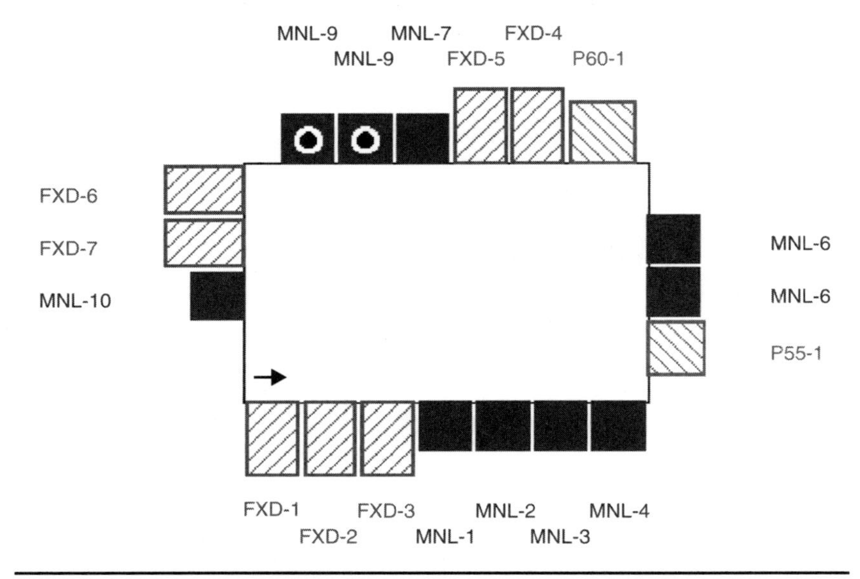

Figure 5.4 Tight loop system #3.

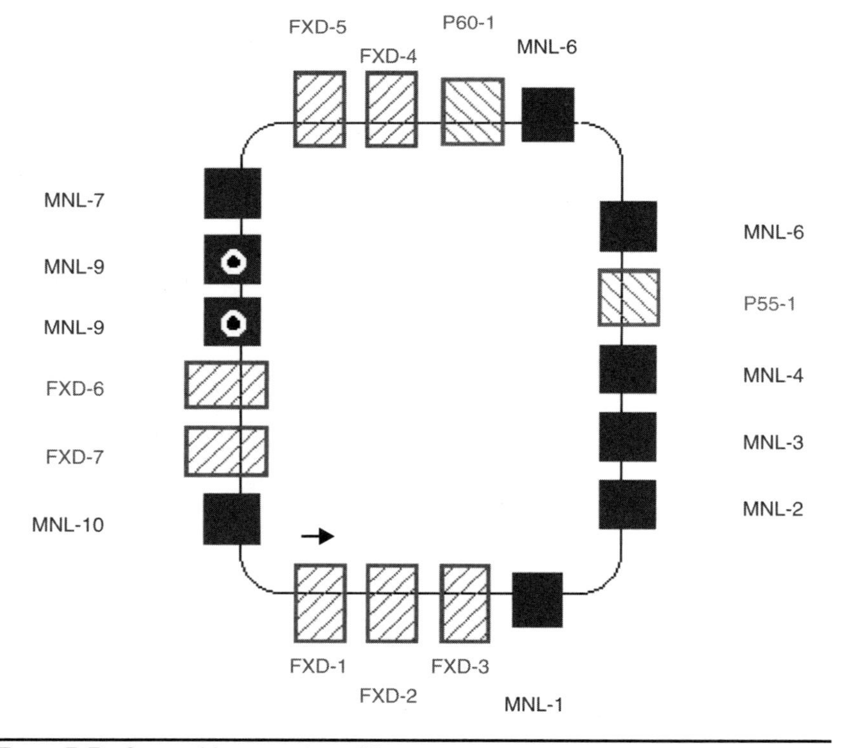

Figure 5.5 Spaced loop system #1.

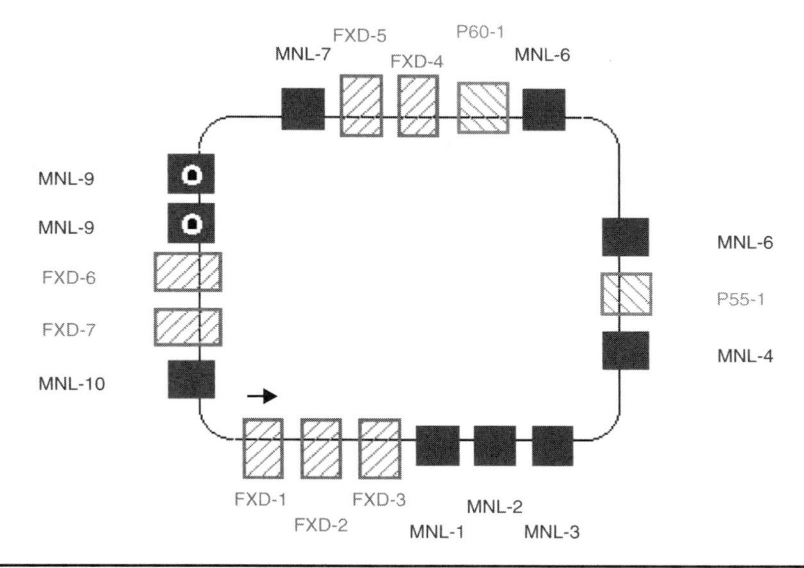

Figure 5.6 Spaced loop system #2.

upon observation of typical industrial practice—one quarter of the specified width of the workstation on either side of it (along the straight sections of the system) is used in these examples. Recall that the goal is to provide an initial system illustration.

5.7 "U" Cell System

The fourth fundamental type of system is composed of workstations arranged in a "U" shape, which has one side totally open. Once again, there can be various shape factors available:

	Number of stations			Dimension ratio (*x*/*y*)
	Sec. 1	Sec. 2	Sec. 3	
Shape 1	5	9	5	0.53
Shape 2	5	8	6	0.65
Shape 3	6	7	6	0.82
Shape 4	6	6	7	1.03
Shape 5	7	5	7	1.32
Shape 6	7	4	8	1.76

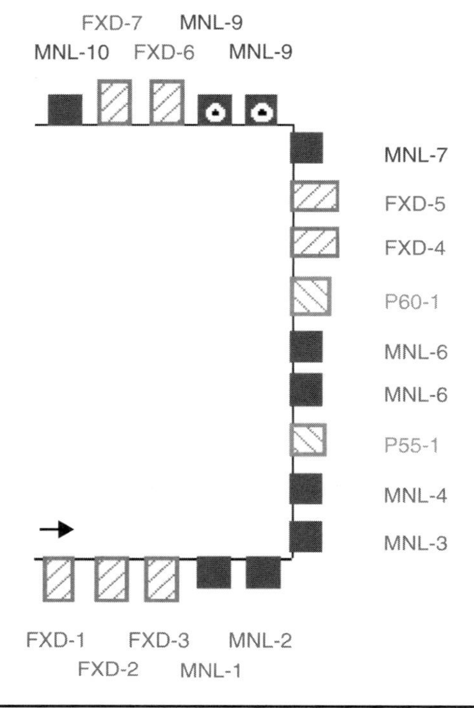

FIGURE 5.7 "U" cell system #1.

Fig. 5.7 shows what a system looks like for the minimum ratio of x-dimension to y-dimension (0.53). The maximum ratio (1.76) of the system is exhibited in Fig. 5.8.

While the "bird's eye" views are interesting, many observers prefer to see a 3-D picture of the system. Two possibilities are presented here: isometric and two-point perspective. The drawings produced by the two techniques will be quite similar for most cases. Rather than exhibit such views for all of the cases shown above, only an example of closed cell with spacing as well as a "U" cell will be shown.

5.8 3-D View of a System

Sometimes very useful insights concerning system behavior can be obtained from an *isometric view*. Example of closed

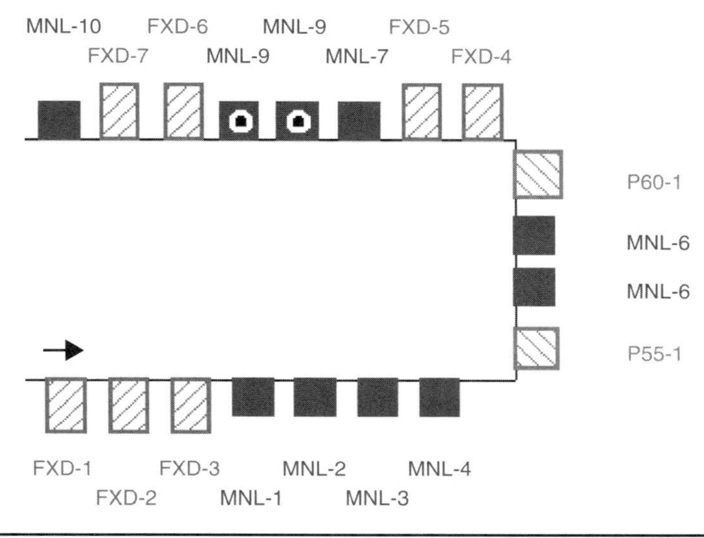

Figure 5.8 "U" cell system #2.

loop with spacing is shown in Fig. 5.9, while an example of "U" cell is shown in Fig. 5.10.

Even better discernment can often be obtained from a perspective view of a system. Example of closed loop with spacing is shown in Fig. 5.11, while an example of "U" cell is shown in Fig. 5.12.

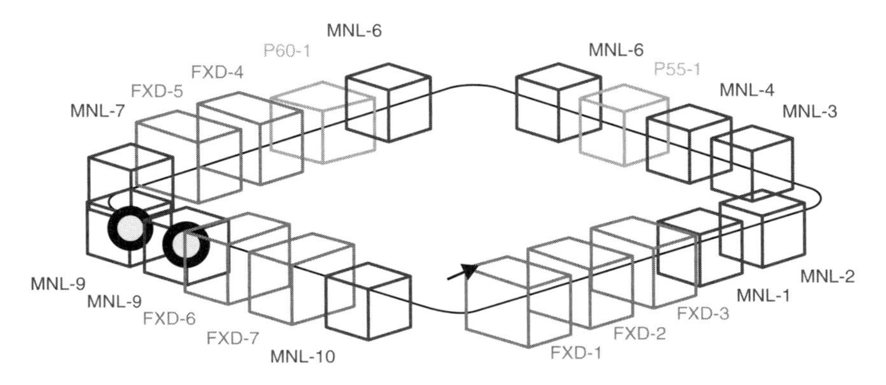

Figure 5.9 Isometric view of spaced loop system.

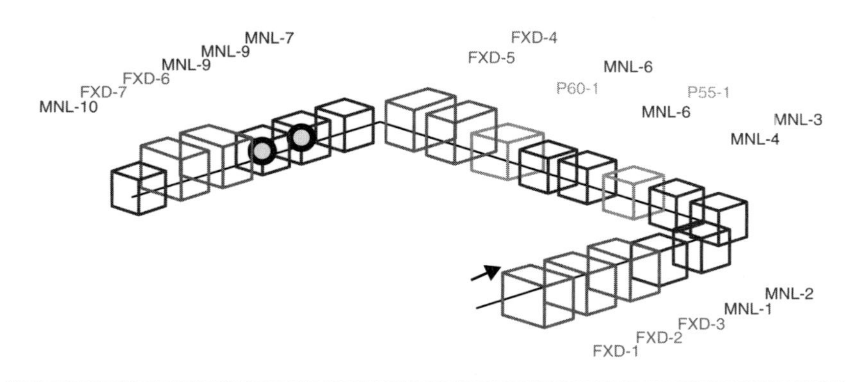

Figure 5.10 Isometric view of "U" cell system.

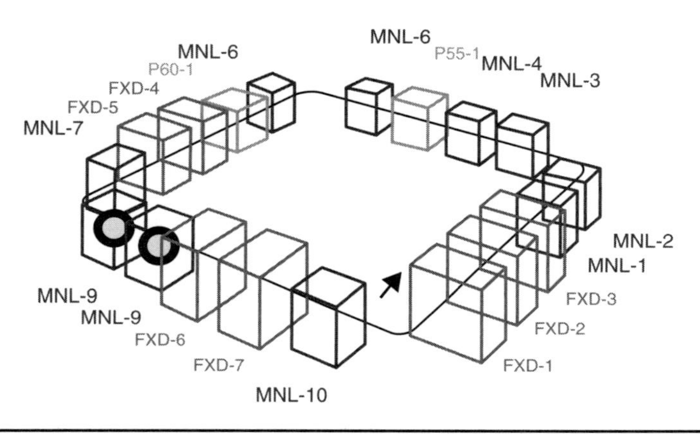

Figure 5.11 Perspective view of spaced loop system.

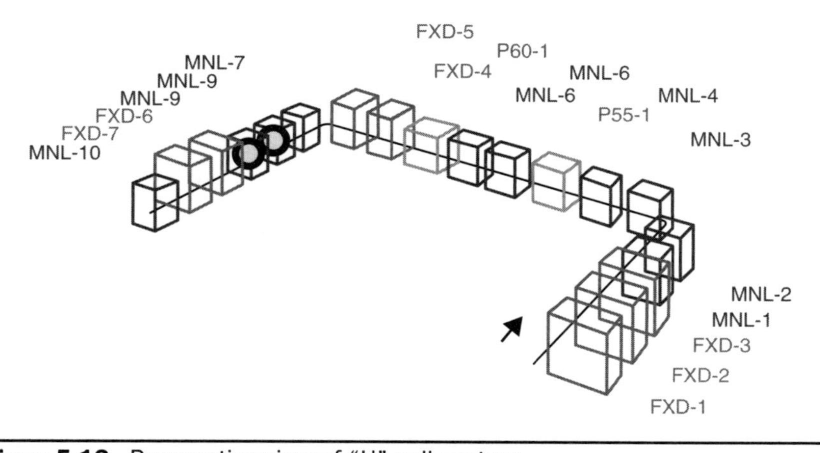

Figure 5.12 Perspective view of "U" cell system.

5.9 Summary

These "first look" pictures provide a very practical sense of how an actual system will be constructed. Generally, there will be a range of shape factors available. No attempt is made here to create accurate pictures for any of the resource types because the goal is to determine the most cost-effective system and only approximately how it would look when implemented.

CHAPTER 6

Multiple Disparate Products Produced by One System

6.1 Introduction

Once a method for determining the best assembly system for a single product has been established using the advanced system design procedure, it is possible to extrapolate the key ideas in order to establish the best single assembly system for a number of products. The fundamental idea is to create systems that will accomplish as many tasks as possible (for each product) at each workstation for the minimum activity-based cost. Chap. 3 contains descriptions of the many parameters that are also used for the multiple-product case.

Product, as used in this chapter, can be an entire device, any portion thereof, various stages of subassembly, or final assembly. The system used as an example is required to perform final assembly of three significant automobile components that have comparable physical characteristics.

Because the procedure can be rapidly utilized, it is valuable at any stage of product/process design. The techniques are applicable to a wide variety of mechanical, electromechanical, and electronic products. Other, totally different and as yet undefined, systems that require unique sets of activities to be performed by a particular set of resources also appear to be manageable by using this design method.

6.2 Fundamental Principles

Economic constraints (e.g., minimum attractive rate-of-return, capital recovery period, nominal wages plus benefits) must be prescribed. The technological limits within which the best assembly system must be determined are essentially the same as those used for a single product. The main constraint on the products that should be considered for assembly on a single system is that they should be similar in size and weight; these parameters along with task degree-of-difficulty are needed to establish the resource type requirements. Although a group of products is to be assembled, only one set of applicable resource types can be specified from the generic categories of manual, fixed automation, and/or programmable automation. Current evidence suggests that it will also be desirable that the products have a similar number of tasks to be performed since the number of unused workstations for any product will likely then be minimized.

There will be a number of usable solutions to the assembly system design problem; each is defined by the maximum workstation time available; usually only one of the products will require the longest time. A method for rating each of the systems by employing user-alterable criteria readily allows them to be ranked and the best one determined. There may be particular (company-imposed) conditions that necessitate choosing a system that has not been top-ranked. Various tables, graphs, and schematics provide specifications for any of the synthesized systems.

6.3 Establishing the Multiple-Product Task/Resource Matrix

The matrix is determined by establishing which resource types (manual, fixed automation, programmable automation) are applicable on a task-by-task basis for each of up to nine different products. One method for doing so is described in Gustavson (1990). In order that a single system be able to assemble all of the products, a shared-resource data set is required. It is not necessary that all resource types be applicable to at least one task for each of the products (as is the case when only a single product is being considered). To keep the spatial requirements of the physical system reasonable, the number-of-tasks limit shown below is used.

Number of products	Maximum tasks for each product
1 or 2	99
3	79
4	59
5	47
6	39
7	33
8	29
9	26

The logistics of multiple-product assembly by a single system, as well as the space available, may provide the practical upper limit to the number of products and/or maximum tasks that can actually be considered. Products that are generally minor variations of each other (e.g., panel meters) constitute a special case of the generic method being discussed and should be handled as a single-product assembly system with "intelligent" workstations.

6.4 Specifying the Production Requirements

General limits on the time available at a workstation are derived from working days per year, maximum shifts available, station-to-station move time, units per pallet, and maximum in-parallel stations. For each product, the following must be specified: production batch size and maximum tools at a station.

6.4.1 Production Batch Size

Production batch size means units to be produced in a work year (month/week/day). When more than one type of assembly is to be produced in a specified period, there must be a rational way to divide that time interval. Allocating time only on production required will usually lead to systems that are far from the best for at least one of the assemblies. The following procedure has been found to be much more useful:

Identify each product by subscript k. Determine, for each resource type, the expected total time for those tasks that can be performed (by the resource type) on each product; call this parameter T_k. With Q_k specified as the production volume, workstation time allotment for each product τ_k will be

$$\tau_k = \frac{Q_k\ T_k}{\sum_j Q_j\ T_j}$$

Total production will be the sum of Q_k, or

$$Q_{\text{total}} = \sum Q_k$$

Dividing the τ_k equation by Q_{total} produces the appropriate decimal portion of total production for each product in any time period. Thus, you can specify system characteristics on the basis of an hourly, daily, weekly, monthly, or yearly production mix. The system, however, will be designed using the equivalent yearly production.

When only two products are involved, the workstation time allocation as a function of production allotment will have the behavior exhibited in Fig. 6.1 for a particular resource type; the convexity or concavity of the curve depends upon the total task time ratio (T_2/T_1).

In general, every resource type in the matrix will have a different ratio. Although up to nine products can be assembled by one system, the behavior for no more than three can be described graphically. Fig. 6.2 shows what happens when a third assembly is added to the two shown in Fig. 6.1; the solid can be thought of as a half-cube. The characteristic for each product is now defined by a smooth-surfaced, irregular volume instead of an area. Note that the data on the left rear of the half-cube is the data shown in Fig. 6.1.

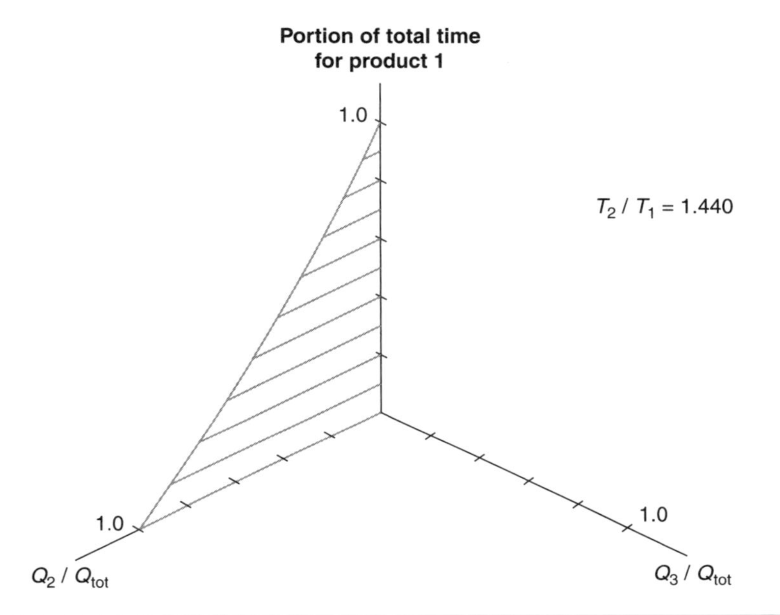

FIGURE 6.1 Workstation time allocation for two products.

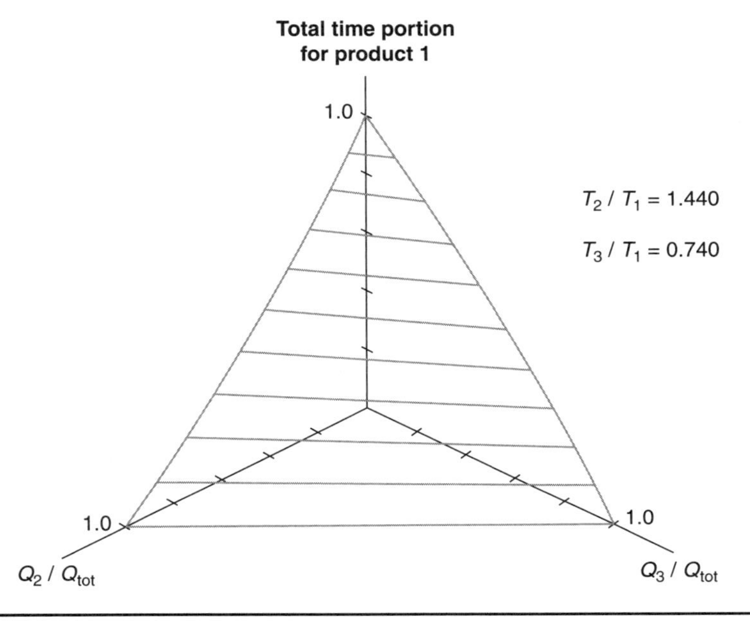

FIGURE 6.2 Workstation time allocation for three products.

6.4.2 Maximum Tools at a Station

The number of maximum tools at a station is governed by the lot size and production requirements. The number of "tools" available at a workstation N_{total} has some physical limit (e.g., 6 for a complex programmable device). Available workstation time (see Sec. 6.3) usually establishes a smaller limit on the number of tools actually used. When the products are to be made in large batches, the maximum may be applied to each product since the whole tooling set will be exchanged simultaneously. For production of very small batches (down to the proverbial "batch-of-one"), the limiting number of tools must be allocated depending upon available time. In general, the maximum number of tools N_k for product k is

$$N_k = \tau_k \times N_{total}$$

Although not often useful for single-product assembly systems, nonconsecutive tasks at any workstation will be allowed for multiple-product systems. This can cause some very complex task assignments for at least one of the products, especially if the number of "skip-over" tasks allowed is greater than 1.

6.5 Determining a Group of Usable Systems

A direct method (synthesis) for finding systems satisfying the technological and economic requirements used for single products (see Chap. 3) has been expanded to include up to nine different products, thus allowing a complete range of production variations. Fundamentally, the goal is to perform as much work as possible for the lowest cost at each workstation regardless of the product mix. The general solution method is as follows:

1. Calculate the time available α_k at a workstation (for each resource type) from

$$\alpha_k = \frac{(T_k/T_1)\,((28800\ S_D\ D_Y\ \varepsilon)/Q_{total})\ F}{\sum_j q_j\ (T_j/T_1)} - t_m$$

where k = product number
F = availability factor: decimal portion of the total time (maximum = 1.0)
28800 = seconds per shift (= 60 s/m × 60 m/h × 8 h/shift)
S_D = shifts per day
D_Y = work days per year
ε = uptime expected (decimal): resource type dependent (<1.0).
t_m = pallet move time = in–out "dead" time.
$q_j = Q_j/Q_{total}$: (production volume portion)

2. Synthesize the most cost-effective system for that availability.
3. Take an amount of time slightly less than that required for the bottleneck station in the prior system (i.e., reduce the availability).
4. Return to step 1 as long as the in-parallel stations constraint is not exceeded.

Each product will have its own combination of unit costs, station compositions, and required investments. The latter includes resource costs that are divided among all products. Fig. 6.3 exhibits the unit cost behavior for each of the three example products to be assembled. Note that the lowest cost for a product occurs for different solutions (which are defined by the availability factor).

Our goal is to determine the "best" system for all three products as a group, not individually (it is possible that each product's most desirable system would be the same). The method for evaluating systems requires that each of them be given a RATING relative to all of the others. Three characteristics (average unit cost, total investment required, and number of stations) are given weighting factors. Desired values for those parameters may also be specified. Each solution has a numerical RATING that provides the basis for a classification of the systems in descending order; the highest ranked solution defines the best system. Fig. 6.4 contains a bar graph and a ranked-listing table showing typical RATING results. Desired values for average unit cost, total investment,

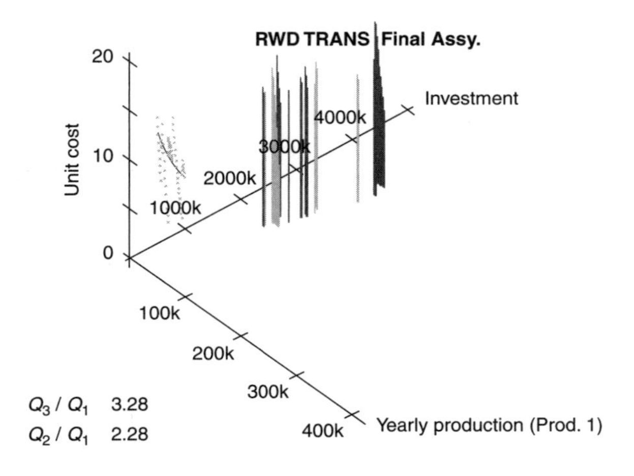

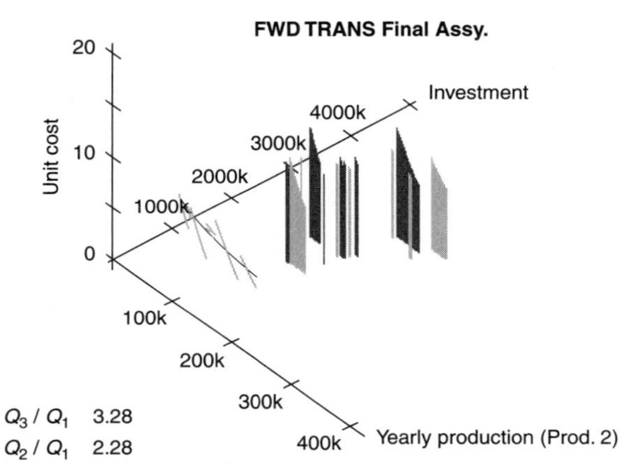

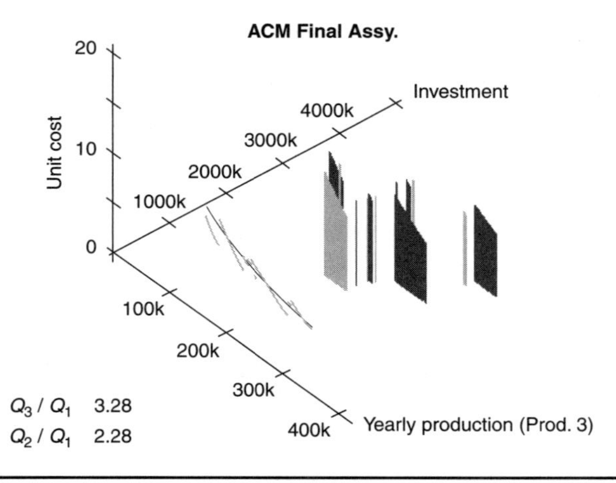

FIGURE 6.3 General solution—unit cost vs. yearly production vs. investment for three 3 products.

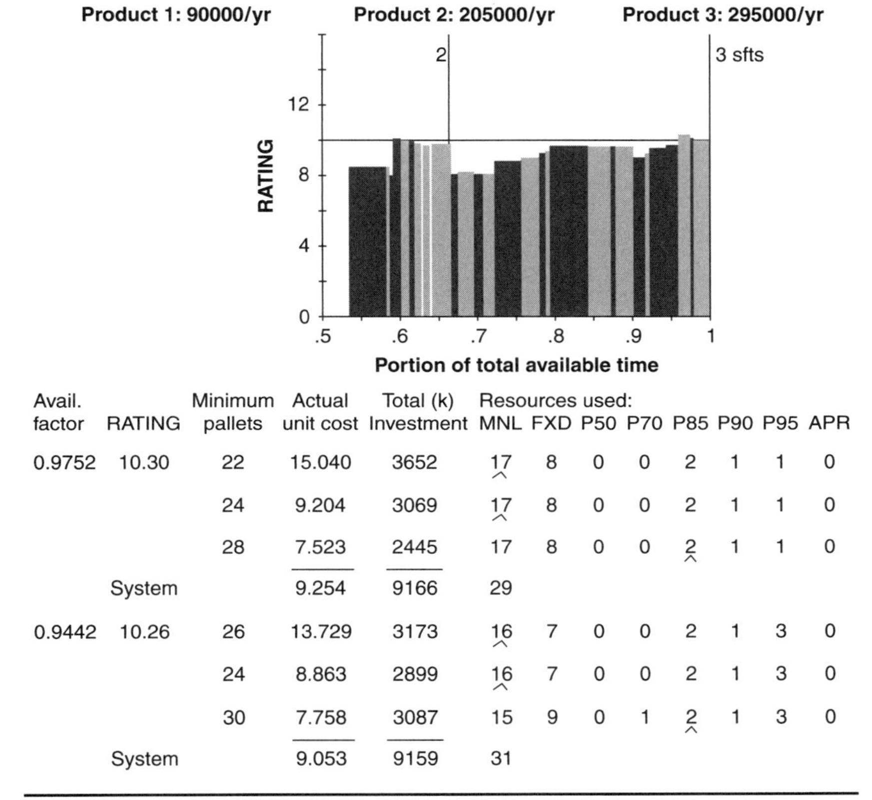

Avail. factor	RATING	Minimum pallets	Actual unit cost	Total (k) Investment	Resources used: MNL	FXD	P50	P70	P85	P90	P95	APR
0.9752	10.30	22	15.040	3652	17	8	0	0	2	1	1	0
		24	9.204	3069	17	8	0	0	2	1	1	0
		28	7.523	2445	17	8	0	0	2	1	1	0
	System		9.254	9166	29							
0.9442	10.26	26	13.729	3173	16	7	0	0	2	1	3	0
		24	8.863	2899	16	7	0	0	2	1	3	0
		30	7.758	3087	15	9	0	1	2	1	3	0
	System		9.053	9159	31							

Figure 6.4 System RATING and ranking (two best systems).

and number of workstations could have been designated. Specifying such values rarely takes place in the early stages of system design since the magnitude of each parameter cannot usually be readily defined. The ranked listing of solutions (Fig. 6.4) is often the most valuable information resulting from the system design process for a particular production requirement. Although the better solutions tend to occur near the shift upper boundaries for systems that contain direct labor, this is not necessarily the case for mostly automated systems.

A decision must be made about whether the best solution is to be used. In the best solution exhibited in Fig. 6.4, for example, products 1 and 2 have at least one MNL (manual) station as their bottleneck while product 3 is controlled by at least one P85 (large programmable) station. The two highest rated systems outlined are very similar; each has a rating slightly better than ideal. Note that they require less than 100% of the available work time (as defined by the availability factor).

6.6 Details of the Best Multiple-Product System

While the following particulars apply to any of the systems found using the general solution procedure, another availability factor (decimal portion of the maximum available time) may be selected, if desired, in order to create these specific details. Table 6.1a exhibits a portion of the cost and performance listing for the best system. After displaying the general constraints, a table of characteristics for each available resource type lists (1) labor cost factor per shift, (2) operating/maintenance factor, and (3) available time (maximum possible for tasks at a workstation) for each product. Note the significant variety of times exhibited in Table 6.1a; this is due to the allocation procedure, described earlier, for each resource type.

Cost and performance details for each workstation of the synthesized system are then accounted for each product. In this system, the first task for product 1, the first two tasks for product 2, and the first task for product 3 are assigned to the first workstation—a manual station (MNL). Annualized resource cost and labor cost (only indirect, for a nonmanual station) have been prorated among the products as per the production volume mix requirements. Labor cost is normally the major portion of the variable cost determined for the first task assigned to any workstation. The tool number (e.g., 101) is assumed to be unique for every product; each of them must buy its own tools.

Workstation definition data is followed by a summary of the system characteristics; see Table 6.1b. For each resource type used in the system, observe total cost, number used, maximum expected workstation time, fixed unit cost, variable unit cost, number of tasks, number of tools, and number of workers. The performance summary identifies units per hour, cycle-time expected, bottleneck station uptime expected, production capacity of this system for the particular

Unused resource available time scale factor = 0.973
233.6 days for 3.0 shifts; 240 days for 2.92 shifts
8.00% min. attr. rate of return; 5-year capital recovery period
22.31% capital eqpt. residual value; 0.00% tooling residual value
24.72 $/h loaded labor rate 3.00-s station-to-station move time
Resource types **[MLTRES-7]**

	MNL	FXD	P50	P70	P85	P90	P95	APR
Hrdwe Cost	200	0	50000	70000	85000	90000	95000	0
rho Factor	1.20	1.00	2.25	2.25	2.63	2.75	2.88	1.00
Uptime %	90.00	95.00	95.00	95.00	95.00	95.00	95.00	95.00
OprMntRate	0.50	1.17	1.00	1.40	1.70	1.80	1.90	1.21
ToolChange	2.00	0.00	5.00	6.00	6.80	7.00	7.30	0.00
Sttn/Wrkr	0.83	5.97	5.00	4.27	3.81	3.67	3.53	100.00
ToolCost %	90.00	35.00	75.00	75.00	75.00	75.00	75.00	35.00
Max Tools	6	1	3	4	6	6	6	1
Years Used	0	0	0	0	0	0	0	0
No Avlble.	299	299	299	299	299	299	299	299
Annl Cost	105	0	26274	36784	52756	58564	64911	0
OprMntFctr	6480	2923	3420	2443	2012	1900	1800	2826
LbrF/Shift	0.0971	0.0135	0.0161	0.0189	0.0212	0.0220	0.0228	0.0008
Product 1 **[TRANTSK8]**; 90000 units/year								
Avltime	36.9	39.2	39.2	39.2	39.2	39.2	39.2	39.2
Product 2 **[TRAXTSK8]**; 205000 units/year								
Avltime	42.5	45.0	45.0	45.0	45.0	45.0	45.0	45.0
Product 3 **[ACM-TSK8]**; 295000 units/year								
Avltime	14.9	15.9	15.9	15.9	15.9	15.9	15.9	15.9

Synthesized system

Task	Resource used	Resource cost	Variable cost	Operation time	Tool change	Tool number	Support cost
1–1	MNL-1	8	26611	25.5	0.0	101-1	1157
2–1	MNL-1	18	60128	15.0	0.0	101-2	1009
2–2	MNL-1	0	789	23.0	2.0	102-2	1009
3–1	MNL-1	26	86457	12.0	0.0	101-3	48116
1–2	P90-1	8949	7018	20.0	0.0	602-1	30566
2–3	P90-1	20333	16268	23.0	0.0	603-2	20038
3–2	P90-1	29282	21569	11.0	0.0	602-3	9509

Table 6.1a Cost and Performance Data for the Best System (Portion)

Product 1 [RWD TRANS. Final Assy.]								
	Total	Number	Time	Unit cost			Number of	
Resource	**cost**	**used**	**used**	**Fixed**	**Variable**	**Tasks**	**Tools**	**Workers**
MNL	534692	17	43.0	0.914	5.027	13	13	20.48
FXD	651512	8	17.3	6.880	0.359	7	7	1.34
P85	71487	2	22.6	0.648	0.146	2	2	0.52
P90	46533	1	24.2	0.439	0.078	1	1	0.27
P95	50332	1	24.2	0.478	0.081	1	1	0.28
Total				**9.359**	**5.692**			**22.90**

83.79 units per hour
43.0 s cycle-time expected
90.00 % bottleneck station uptime expected
System capacity: bottleneck, 93767
Triangular Dist.: RN best, 93767; RN worst, 90678
Exponential Dist.: RN best, 93767; RN worst, 90016
24.96 $/h system operating/maintenance rate
1353600 cost ($) to produce 90000 units; with unit cost ($) 15.040
System annualized charge factor 0.2306
3651974 ($) total investment required
3230968 ($) for required hardware
1904513 ($) capital equipment
1326455 ($) tooling

Product 2 [FWD TRANS. Final Assy.]								
	Total	Number	Time	Unit cost			Number of	
Resource	**cost**	**used**	**used**	**Fixed**	**Variable**	**Tasks**	**Tools**	**Workers**
MNL	1196458	17	49.7	0.799	5.052	13	13	20.48
FXD	482080	8	20.5	2.000	0.357	7	7	1.34
P85	99088	2	30.5	0.336	0.148	2	2	0.52
P90	56639	1	27.4	0.197	0.080	1	1	0.27
P95	51481	1	26.3	0.169	0.082	1	1	0.28
Total				**3.502**	**5.720**			**22.90**

72.40 units per hour
49.7-s cycle-time expected
90.00% bottleneck station uptime expected
System capacity: bottleneck 213581
Triangular Dist.: RN best, 213581; RN worst, 203351
Exponential Dist.: RN best, 213581; RN worst, 197665
24.96 $/h system operating/maintenance rate
1886820 cost ($) to produce 205000 units, with unit cost ($) 9.204
System annualized charge factor 0.2333
3069376 ($) total investment required
2581929 ($) for required hardware
1381269 ($) capital equipment
1200660 ($) tooling

TABLE 6.1b Cost and Performance Summary for the Best System (Portion)

Resource-types data: **mltres-7** Tasks data:
(1) trantsk8 (2) traxtsk8 (3) acm-tsk8

42.96 seconds Usable cycle time
90.00 % Bottleneck up-time 83.79 Units/hr expected

System capacity: bottleneck 93767
Triangular Dist.: RN best 93767 RN worst 90678
Exponential Dist.: RN best 93767 RN worst 90016

3651974 ($) Total investment, rho factor = 1.13
1904416 ($) Capital equipment 1326455 ($) tooling

22.90 Workers at 24.72 $/hr required
24.96 $/hr System operating/maintenance rate

0.186 Year required for 3.0 shift operation
240 Days required for 0.56 shift operation

90000 units/yr **$15.040 Each** **0.975 AF**

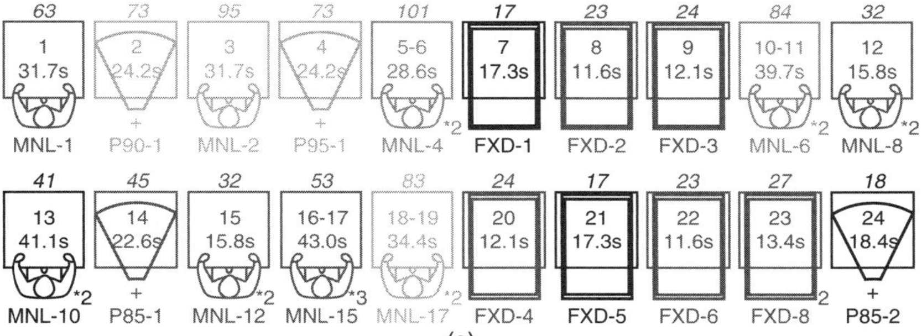

63 1 31.7s MNL-1
73 2 24.2s + P90-1
95 3 31.7s MNL-2
73 4 24.2s + P95-1
101 5-6 28.6s *2 MNL-4
17 7 17.3s FXD-1
23 8 11.6s FXD-2
24 9 12.1s FXD-3
84 10-11 39.7s *2 MNL-6
32 12 15.8s *2 MNL-8
41 13 41.1s *2 MNL-10
45 14 22.6s + P85-1
32 15 15.8s *2 MNL-12
53 16-17 43.0s *3 MNL-15
83 18-19 34.4s *2 MNL-17
24 20 12.1s FXD-4
17 21 17.3s FXD-5
23 22 11.6s FXD-6
27 23 13.4s 2 FXD-8
18 24 18.4s + P85-2

(a)

Figure 6.5 Schematic layout for the best system. (a) Product 1; (b) product 2, and (c) product 3. (*Continued*)

product, system operating/maintenance rate, unit cost, system annualized charge factor, total investment required, total hardware cost, total for capital equipment, and total for tooling (support equipment). Note that labor and resource cost are prorated.

Important information can be found in a schematic layout for any manufacturing system such as those shown in Figs. 6.5a, 6.5b, and 6.5c, respectively, for the three products used in the example. Color for each station can represent the highest degree-of-difficulty of a task to be performed there: blue, easy; green, moderate; yellow, complex; red, must be done

Resource-types data: **mltres-7** Tasks data:
(1) trantsk8 **(2) traxtsk8** (3) acm-tsk8

49.72 seconds Usable cycle time
90.00 % Bottleneck up-time 72.40 Units/hr expected

System capacity: bottleneck 213581
Triangular Dist.: RN best 213581 RN worst 203351
Exponential Dist.: RN best 213581 RN worst 197665

3069376 ($) Total investment, rho factor = 1.19
1381269 ($) Capital equipment 1200660 ($) tooling

22.90 Workers at 24.72 $/hr required
24.96 $/hr System operating/maintenance rate

0.492 Year required for 3.0 shift operation
240 Days required for 1.47 shift operation

205000 units/yr **$9.204 Each** **0.975 AF**

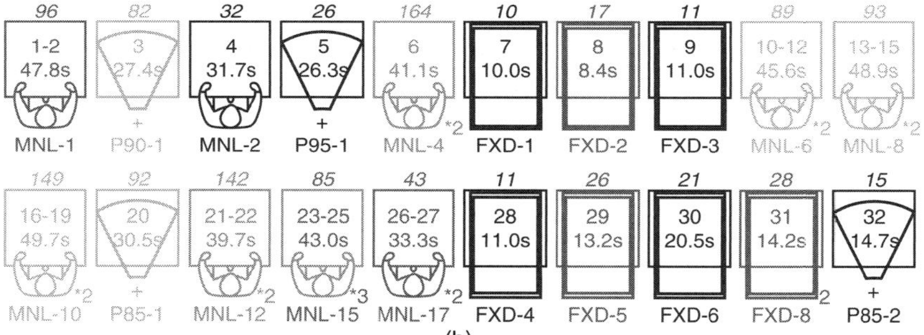

96 1-2 47.8s MNL-1; 82 3 27.4s + P90-1; 32 4 31.7s MNL-2; 26 5 26.3s + P95-1; 164 6 41.1s *2 MNL-4; 10 7 10.0s FXD-1; 17 8 8.4s FXD-2; 11 9 11.0s FXD-3; 89 10-12 45.6s *2 MNL-6; 93 13-15 48.9s *2 MNL-8

149 16-19 49.7s *2 MNL-10; 92 20 30.5s + P85-1; 142 21-22 39.7s *2 MNL-12; 85 23-25 43.0s *3 MNL-15; 43 26-27 33.3s *2 MNL-17; 11 28 11.0s FXD-4; 26 29 13.2s FXD-5; 21 30 20.5s FXD-6; 28 31 14.2s 2 FXD-8; 15 32 14.7s + P85-2

(b)

Figure 6.5 *(Continued)*

manually. Note that each product has a different bottleneck station. Most of the stations with at least one very difficult task (the "red" ones) have expected throughput times that are 90%, or less, of the respective product cycle-time; they effectively each have a time buffer built-in to their performance requirements.

Product	Figure	Bottleneck station	Most difficult task
1	6.5(a)	14 (MNL-15)	moderate
2	6.5(b)	11 (MNL-10)	complex
3	6.5(c)	12 (P85-1)	complex

Resource-types data: **mltres-7** Tasks data:
(1) trantsk8 (2) traxtsk8 **(3) acm-tsk8**

18.42 seconds Usable cycle time
95.00 % Bottleneck up-time 195.43 Units/hr expected

System capacity: bottleneck 307349
Triangular Dist.: RN best 303476 RN worst 297858
Exponential Dist.: RN best 307349 RN worst 287972

2445230 ($) Total investment, rho factor = 1.20
1329670 ($) Capital equipment 703230 ($) tooling

22.90 Workers at 24.72 $/hr required
24.96 $/hr System operating/maintenance rate

0.262 Year required for 3.0 shift operation
240 Days required for 0.79 shift operation

295000 units/yr **$7.523 Each** **0.975 AF**

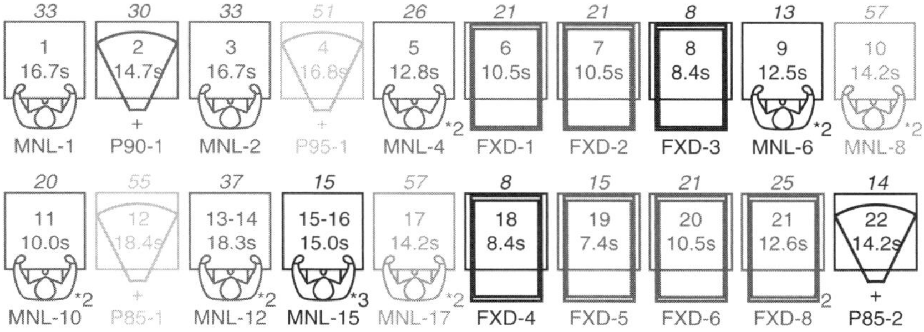

(c)

Figure 6.5 *(Continued)*

Sensitivity of cost for variations in production volume can be easily established (Gustavson, 1996). When the production volumes are to be changed significantly, the whole system design process should be repeated using those new values. Most manufacturing systems are the best for only a very limited range (plus or minus a few percent) of the prescribed production volumes.

6.7 Management Overview of a System

This view of the system requires cost and performance data from the synthesized system as well as the task descriptions

Required annual capacity 90000/year for product 1
Resource-types data: **MLTRES-7;** tasks data: **TRANTSK8**

Station number	Station type	Description	Effective capacity (Unit/h)	Capital equipment cost	Tooling cost	Total cost
100	MNL-1	Attach main housing—c3, c4, p1 ring assy. (A) to pallet	113	0 k$[†]	4 k$	4 k$
200	P90-1	Assemble P1 carrier–P2 ring Assy. (G)	148	25k$	34k$	59k$
300	MNL-2	Assemble clutch stack (K)	113	0k$	4k$	4k$
400	P95-1	Assemble P2 carrier–P3 ring Assy. (H)	148	26k$	35k$	61k$
500	MNL-4*2	Assemble main shaft (J)	125	1k$	9k$	10k$
		Align rear housing—output shaft Assy. (L)		1k$	10k$	11k$
600	FXD-1	Insert bolts	207	90k$	40k$	139k$
700	FXD-2	Torque bolts	310	1k$	4k$	4k$
800	FXD-3	Test assembled components	297	57k$	6k$	64k$
		. . .				
1200	P85-1	Torque bolts.	159	22k$	27k$	49k$

1300	MNL-12*2	Align bell housing (D)	227	1k$	7k$	8k$
1400	MNL-15*3	Insert bolts.	83	1k$	7k$	8k$
		Torque bolts		19k$	168k$	187k$
1500	MNL-17*2	Assemble torque converter assembly (E)	104	1k$	9k$	10k$
		Position end bolt and seal (M) and tighten fasteners		1k$	5k$	5k$
1600	FXD-4	Align front cover (F)	297	163k$	88k$	252k$
1700	FXD-5	Insert bolts	207	90k$	49k$	139k$
1800	FXD-6	Torque bolts	310	168k$	91k$	259k$
1900	FXD-8*2	Perform final test	268	981k$	528k$	1509k$
2000	P85-2	Pack/unload assembly	195	20k$	22k$	42k$
Total				**1905k$**	**1326k$**	**3231k$**

† k$ = thousand dollars.

Table 6.2a Management View of the Best System (Product 1, Portion)

Required annual capacity 204480/year for product 2 Resource-types data: **MLTRES-7;** tasks data: **TRAXTSK8**						
Station number	**Station type**	**Description**	**Effective capacity (Unit/h)**	**Capital equipment cost**	**Tooling cost**	**Total cost**
100	MNL-1	Attach MIN CASE (O) to pallet	75	0k$[†]	3k$	3k$
		Assemble R/L clutch piston spring Assy.; snap ring; Pr. plate (C)		0k$	3k$	3k$
200	P90-1	Assemble R/L ONC outer; R/L ONC inner; snap ring (X)	131	39k$	22k$	61k$
300	MNL-2	Rotate trunnion 180°	113	0k$	3k$	3k$
400	P95-1	Snap fit O-ring & fluid filter; lube tube (D) into assembly	136	37k$	13k$	50k$
500	MNL-4*2	Assemble MLCS Cm/Lvr; Shft; SnpRng; Park Pawl; Etc. (H)	87	1k$	9k$	10k$
600	FXD-1	Place band (M) into position	360	78k$	42k$	120k$
700	FXD-2	Align oil pump; gasket (A)	427	177k$	95k$	272k$
800	FXD-3	Insert bolts	326	83k$	45k$	128k$
900	MNL-6*2	Torque bolts	79	16k$	144k$	161k$
		Rotate trunnion 180°		1k$	5k$	5k$
		Assemble triple clutch pack; pump end washer (U)		1k$	6k$	7k$

1000	MNL-8*2	Snap fit Servo; measure depth; springs; retaining ring. (G)	73	1k$	4k$	5k$
		Assemble turbine shaft; support end washer; RTB (R)		1k$	7k$	8k$
		Assemble OWC; L/I SUN ASSY.; RTB (S)		1k$	6k$	7k$
		. . .				
1500	MNL-17*2	Position filler tube; V.S.S.; speedo (I) and tighten fasteners	108	1k$	5k$	5k$
		Align main control; gasket; slave pins (B)		1k$	6k$	7k$
1600	FXD-4	Insert bolts	326	85k$	46k$	130k$
1700	FXD-5	Torque bolts	273	196k$	105k$	301k$
1800	FXD-6	Torque bolts	175	124k$	66k$	190k$
1900	FXD-8*2	Perform final test	253	436k$	235k$	671k$
2000	P85-2	Pack/unload assembly	244	35k$	16k$	52k$
Total				**1382k$**	**1201k$**	**2582k$**

[†] k$ = thousand dollars.

TABLE 6.2b Management View of the Best System (Product 2, Portion)

Required annual capacity 294480/year for product 3 Resource-types data: **MLTRES-7;** tasks data: **ACM-TSK8**						
Station number	**Station type**	**Description**	**Effective capacity (Unit/h)**	**Capital equipment cost**	**Tooling cost**	**Total cost**
100	MNL-1	Attach Al case; Al valve; vacuum element; link (P) to pallet	215	0k$[†]	3k$	4k$
200	P90-1	Install MVH sub-Assy. (B)	244	48k$	10k$	59k$
300	MNL-2	Snap fit evaporator case (L) into assembly	215	0k$	3k$	4k$
400	P95-1	Assemble temperature valve (S)	213	51k$	11k$	62k$
500	MNL-4*2	Position temp. valve actuator (F) and tighten fasteners	281	1k$	7k$	8k$
600	FXD-1	Position solenoid #1 (O) and tighten fasteners	342	164k$	89k$	253k$
700	FXD-2	Snap fit vacuum element #2 (E) into assembly	342	164k$	89k$	253k$
800	FXD-3	Place motor-to-case sealant (C) into position	427	54k$	29k$	83k$
900	MNL-6*2	Place motor & fan; isolator (H) into position	288	1k$	4k$	5k$
1000	MNL-8*2	Snap fit harness (U) into assembly	254	1k$	9k$	10k$
1100	MNL-10*2	Test assembled components	360	1k$	6k$	7k$
1200	P85-1	Orient 3 Stud Seals (M)	195	46k$	10k$	56k$

1300	MNL-12*2	Position resistor Assy. (K) and tighten fasteners	196	1k$	6k$	7k$
		Position solenoid Assy. (T) and tighten fasteners		0k$	0k$	0k$
1400	MNL-15*3	Place evaporator core sub-Assy. (N) into position	240	1k$	6k$	7k$
		Place heater core; heater core shroud; clamp (R) into position		0k$	0k$	0k$
1500	MNL-17*2	Assemble pipe seal (J)	254	1k$	9k$	10k$
1600	FXD-4	Place cover-to-case sealant (G) into position	427	54k$	29k$	84k$
1700	FXD-5	Align cover (A)	488	85k$	46k$	130k$
1800	FXD-6	Torque bolts	342	148k$	80k$	228k$
1900	FXD-8*2	Perform final test	268	461k$	248k$	709k$
2000	P85-2	Pack/unload assembly	253	45k$	9k$	54k$
Total				**1330k$**	**703k$**	**2033k$**

† k$ = thousand dollars.

TABLE 6.2c Management View of the Best System (Product 3)

Resource-types data: **MLTRES-7** Tasks data:
(1) **TRANTSK8** (2) **TRAXTSK8** (3) **ACM-TSK8**

90000 units/yr Prod. 1:	BN 93767	T+ 93767	T– 90678	E+ 93767	E– 90016
204480 units/yr Prod.2:	BN 213581	T+ 213581	T– 203351	E+ 213581	E– 197665
294480 units/yr Prod.3:	BN 307349	T+ 303476	T– 297858	E+ 307349	E– 287972

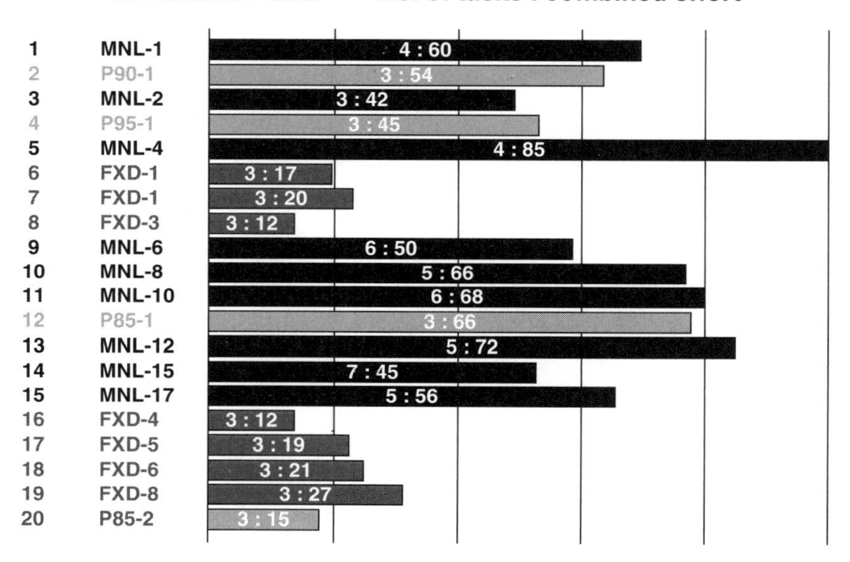

Figure 6.6 Summary effort required at each workstation.

from the assembly process plans. After specifying a workstation numbering increment, you can exhibit (significant portions are shown in Table 6.2) station number, resource type allocated with system usage counter, task description, effective capacity, capital equipment cost, tooling cost, and total hardware cost for each product. Inspection of this data can provide a basis for management requests for changes in the characteristics of the system to modify the conditions for one or more products. This may lead to higher costs; but at least you will now be able to rapidly show how and why that occurred.

Another way to observe the characteristics of a multiproduct system is by looking at the total number of tasks to be performed and the effort required. The necessary effort is defined as the sum of the task times multiplied by the corresponding degree-of-difficulty. Fig. 6.6 exhibits the information for the system just delineated.

6.8 Summary

For a production system where the tasks for up to nine similar products and one corresponding set of applicable resource types can be specified, it is now relatively easy to determine a number of usable systems from which the best can be readily established. All of the necessary details are available in graphical or tabular form.

CHAPTER 7

World-Class Versus Mostly Manual Systems

7.1 Introduction

Although many system types have been and can be designed using the methods described in this book, optimum behavior can only be expected from them as long as the production and cost requirements are reasonably consistent. Other behavior will probably limit the utility of such a system to very narrow circumstances, for example,

1. a new-to-market product may have significant fluctuations in demand,
2. the time when it is viable may be shorter or longer than expected, and/or
3. yearly cost changes might be significant.

In this chapter, two extreme cases will be compared. Some manufacturers will also investigate intermediate cost and capability systems when a long product life and/or significant investment will be required. The two system types that will be used as the basis for investigation are the following:

1. **Mostly manual system** (MM)—most assembly system managers are very comfortable with such an arrangement, particularly since it is often essentially how they presently operate. Limited automation is possible.
2. **World-class system** (WC)—composed of the most cost-effective combination of people, fixed automation, and programmable automation.

Either type can be determined using the methods described in Chap. 3. The design process assumes starting with a "blank sheet of paper."

In this chapter, the cost improvement to be expected when a WC system is to be compared to a MM system will be established. Note that the comparison being made here requires that all of the investment in either system takes place in year zero. Table 7.1 exhibits constant cost and performance characteristics of both the systems for an example product. The economic justification techniques explained in App. A will be used for the comparison with one major inversion: In App. A, the internal rate-of-return (IRoR) is specified and an allowable investment is determined. In this chapter, the actual investment is specified and the IRoR is calculated. Specific data will be obtained by subtracting the appropriate MM value from the WC value (see Table 7.1).

1. Incremental system cost: 2558220 – 595820 = 1962400
2. Incremental major equipment cost: 1093345 – 308435 = 784910
3. Incremental minor equipment cost: 475755 – 263665 = 212090

From these results, the following system comparison parameters can be calculated:

1. Incremental hardware cost: 784910 + 212090 = 997000

	Base	Alternative
Economic data file name	ACM395M4	ACM395W4
Resource type file name	acm-resm	acm-res
Task data file name	acm-tskm	acm-tsk
Direct workers per shift	10.00	2.00
Indirect workers per shift	2.38	2.63
Average loaded labor rate (per hour)	25.90	25.90
Operating/maintenance rate (per hour)	7.34	14.48
Expected system cycle-time (second)	50.92	50.98
Total system cost	595820	2558220
System rho factor	1.041	1.630
Major equipment cost	308435	1093345
Minor equipment cost	263665	474755

	Major equipment depreciation	Minor equipment depreciation	Tax rate
1	14.29%	33.33%	34.00%
2	24.29%	44.44%	34.00%
3	17.49%	14.81%	34.00%
4	12.49%	7.41%	34.00%
5	8.92%	0.00%	34.00%
5*	22.31%	0.00%	

* Salvage value.

Table 7.1 Example Competing System Specifications—Constant Data

2. Incremental minor equipment portion: 212090/997000 = 21.27%
3. Incremental depreciable portion: 997000/1962400 = 50.81%

All six of these values are fixed for information shown in the subsequent sections of this chapter.

While any system is designed for particular output requirements over a number of years (also termed the capital recovery period), the actual performance may be quite different. Does that make an appreciable difference?

7.2 The Constant Value Situation

Since the manufacturing system has been designed using a particular set of economic and time constraints (along with limitations on the applicable resource types), the current values for

1. minimum attractive rate-of-return (MARR)
2. capital recovery period
3. final year % salvage value
4. average hourly wage for year 1
5. station-to-station move time
6. working days per year
7. shifts available

must be known. These nonvarying characteristics will be used as the basis for comparing the alternative WC system to the base MM system. As described in Chap. 3, changing any of those primary constraints usually has fundamental consequences for the design of the resulting system.

For the example ACM Final Assembly, the best systems for a particular yearly production volume have been determined. Comparison of the systems found using the methods described in this book is usually a straightforward task. The easiest comparison is based upon the data for an MM system and a WC system whose unchanging characteristics are shown in Table 7.1.

Fig. 7.1 exhibits annual savings for the constant cost and production volume case when using a WC system instead of an MM system for a range of ACM Final Assembly yearly output. Note that each such comparison applies to a normally narrow range around the particular system production required. For example, the WC and MM systems for

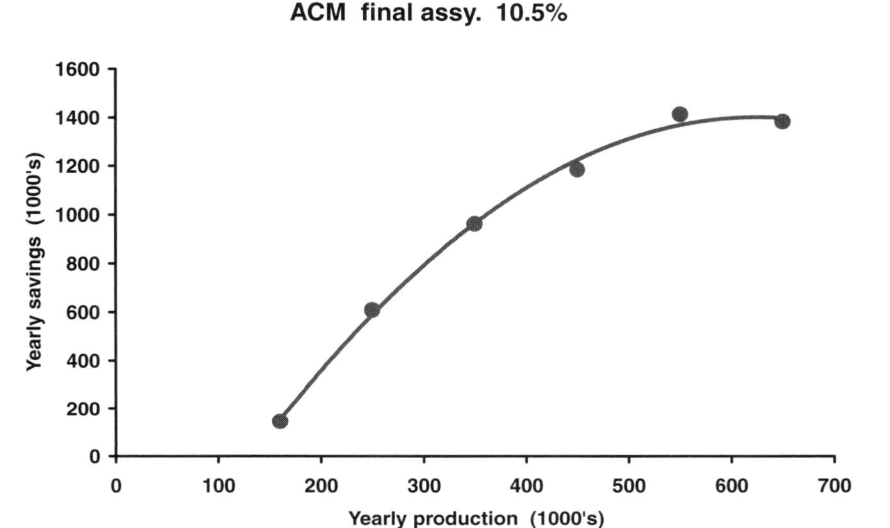

Figure 7.1 Yearly WC vs. MM savings characteristic for a typical product.

250000 units are quite different from the WC and MM systems required for 350000 units.

Seeing how reasonably consistent the annual savings vs. yearly production is, you might conclude that the unit cost improvement would have similar characteristics. Unfortunately, that is not the case. Fig. 7.2 shows the significant variations that occur for year 1 through year 5 over the range of production requirements. The important point to remember is that the procedures described in this book rapidly allow determination of the best system for any combination of tasks, resources types, economic requirements, and time constraints.

When production volume is constant and costs do not change (assumptions made during the system design procedure), the resultant behavior is similar to that shown in Table 7.2 and Fig. 7.3. In the table and figure you will notice that for any full year after year 1, the actual IRoR exceeds the minimum attractive rate-of-return (MARR). Thus, for the constant cost and production case, it is highly likely that the company will obtain significant profit benefits from using the WC system. Not all comparisons will be as optimistic as

	Base	Alternative
Year 1 variable unit cost	4.744	1.872
Year 2 variable unit cost	4.744	1.872
Year 3 variable unit cost	4.744	1.872
Year 4 variable unit cost	4.744	1.872
Year 5 variable unit cost	4.744	1.872

	Units	Yearly savings	IRoR	Unit cost improvement
1	395000	1134677	−0.55%	−0.007
2	395000	1134677	20.00%	0.228
3	395000	1134677	31.00%	0.324
4	395000	1134677	37.60%	0.372
5	395000	1134677	41.86%	0.394

Table 7.2 Example Competing System Characteristics—Constant Data

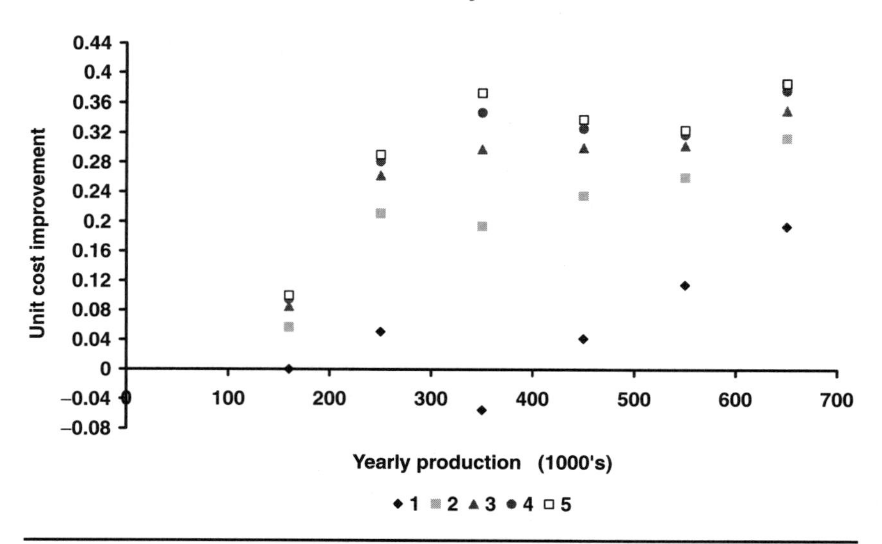

Figure 7.2 Yearly WC vs. MM unit cost improvement for a typical product.

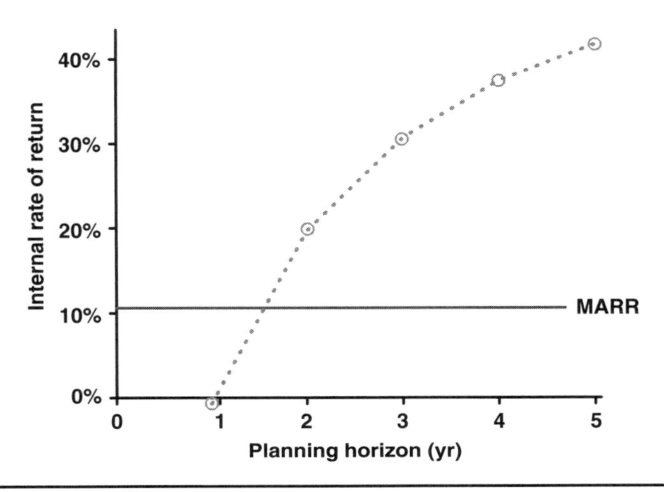

Figure 7.3 Yearly IRoR behavior for constant costs and output.

this one; the numbers must be properly determined for each situation.

7.3 Nonconstant Yearly Costs

The first alternative to be investigated involves changing the cost basis for each year of the capital recovery period while keeping production volume constant. Several possible scenarios include

1. constant percentage increase in costs (inflation)
2. constant percentage decrease in costs (deflation)
3. up and down changes in costs (roller coaster)
4. yearly increase in costs follows a mathematical curve

Any of these conditions, or some other, could be used as the basis for comparing the WC system to the MM system. Here, the constant percentage increase situation will be used. As shown in Table 7.3, there will be an improvement in IRoR in all years of the model. In other words, yearly cost increases produce greater expected profits.

	Base	Alternative
Year 1 variable unit cost	4.744	1.872
Year 2 variable unit cost	4.910	1.937
Year 3 variable unit cost	5.082	2.005
Year 4 variable unit cost	5.260	2.075
Year 5 variable unit cost	5.444	2.148

	Units	Yearly savings	IRoR	Average unit cost improvement
1	395000	1134677	−0.55%	−0.007
2	395000	1174390	20.65%	0.251
3	395000	1215494	32.45%	0.3674
4	395000	1258036	39.75%	0.438
5	395000	1302068	44.40%	0.494

TABLE 7.3 Example Competing System Characteristics—Increasing Costs

7.4 Changes in Yearly Production Volume

A second possibility is that while costs remain constant the production volume changes yearly. Several available types include

1. constant percentage increase in production
2. constant percentage decrease in production
3. production based on a parabolic curve

In all years, production cannot exceed the capacity of the WC or MM system, whichever is lowest. When output increases at a constant percentage, the results for the case at hand are shown in Table 7.4. Since the production volume (and therefore, the expected cost savings) in every year except the fifth is less than that exhibited in Table 7.3 and Fig. 7.3, you should expect to see lower IRoR for each group of years, and that's indeed the case. However, the fundamental

	Base	Alternative
Year 1 variable unit cost	4.744	1.872
Year 2 variable unit cost	4.744	1.872
Year 3 variable unit cost	4.744	1.872
Year 4 variable unit cost	4.744	1.872
Year 5 variable unit cost	4.744	1.872

	Units	Yearly savings	IRoR	Average unit cost improvement
1	330000	947958	−8.09%	−0.126
2	346500	995356	12.09%	0.170
3	363825	1045123	23.68%	0.249
4	382016	1097380	31.18%	0.377
5	401117	1152248	36.19%	0.432

Table 7.4 Example Competing System Characteristics—Increasing Output

desirability of the WC system has not been economically compromised; IRoR still exceeds MARR after year 1.

The second changing yearly production volume case to be evaluated involves what is often called "parabolic production." This could well be the situation when a product is expected to have a finite life during which customer requirements start to decrease (sometimes precipitously) after the maximum system output is reached. A further wrinkle is to have production ramp-up during the early years. Table 7.5 and Fig. 7.4 show how such behavior could occur. The yearly IRoR continues to be greater than the MARR!

7.5 Changes in Yearly Costs and Production Volume

Adding increasing yearly costs to the parabolic yearly production volume example of Table 7.5 produces the data seen in Table 7.6. The savings will be greater each year, compared to data in Table 7.5, and therefore the IRoR will also be greater.

	Base	Alternative
Year 1 variable unit cost	4.744	1.872
Year 2 variable unit cost	4.744	1.872
Year 3 variable unit cost	4.744	1.872
Year 4 variable unit cost	4.744	1.872
Year 5 variable unit cost	4.744	1.872

	Units	Yearly savings	IRoR	Average unit cost improvement
1	360000	1034136	−4.61%	−0.066
2	395000	1134677	17.27%	0.220
3	405000	1163403	29.14%	0.340
4	335000	1019773	35.20%	0.309
5	275000	789965	38.15%	0.260

Table 7.5 Example Competing System Characteristics—Parabolic Output with Constant Yearly Costs

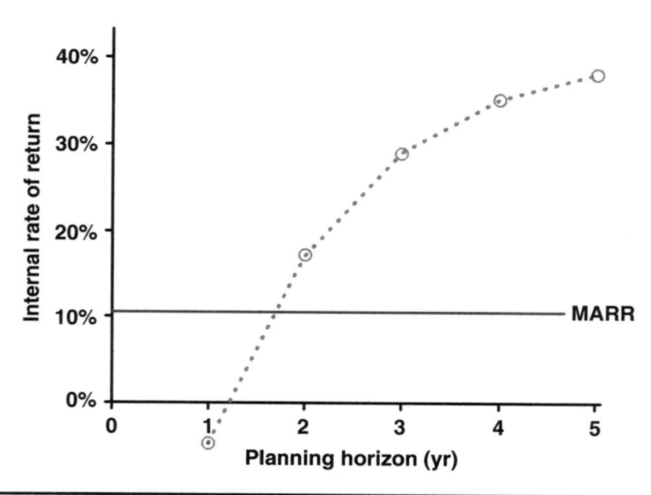

Figure 7.4 Yearly IRoR behavior for constant costs but parabolic output.

	Base	Alternative
Year 1 variable unit cost	4.744	1.872
Year 2 variable unit cost	4.910	1.937
Year 3 variable unit cost	5.082	2.005
Year 4 variable unit cost	5.260	2.075
Year 5 variable unit cost	5.444	2.148

	Units	Yearly savings	IRoR	Average unit cost improvement
1	360000	1034136	−4.61%	−0.066
2	395000	1174390	18.12%	0.233
3	405000	1246266	30.75%	0.375
4	335000	1130640	37.34%	0.422
5	275000	906503	40.58%	0.354

Table 7.6 Example Competing System Characteristics—Parabolic Output with Increasing Yearly Costs

7.6 Summary

For any system designed using the economic–technological synthesis methods described earlier in this book, the techniques explained in this chapter provide a straightforward method for demonstration to management that the company will make at least the minimum attractive rate-of-return for almost all future possibilities. Initial rational choices for the economic and time constraints are assumed to have been made.

The situation that usually causes a problem is due to a 1-year capital recovery period. If that constraint must be used, the MARR generally must be less than about 5%, if any automation is to become part of the system.

APPENDIX A

Determining Allowable Investment

A.1 Introduction

Before understanding how a cost-effective system can be created, a method for determining the economic basis for the decision must be defined. Usually, a comparison is made between a base (either currently installed or proposed) system and an alternative. Corporate strategic considerations are external to the analysis described in this appendix. The method used here is described in Dorf (1988) and Nevins and Whitney (1989); it inverts the typical economic justification analysis found in engineering economy textbooks (Kurtz, 1984).

A.2 Description of a New Technique

While material costs and institutional costs (sales, general and administrative, etc.) are often significant, they usually will not contribute to the comparison of manufacturing system options. The two system cost types of concern are fixed and variable:

Fixed cost—annualization of the initial investment ("rho factor" times the hardware cost) over a capital recovery period in years.

Variable cost—a combination of the loaded labor rate, number of workers (direct and indirect), operating/maintenance rate, and the effective system cycle-time.

When an alternative system can provide reduced variable costs, its fixed cost may be justifiable. How is that condition established? There is currently no single criterion used by every organization; the method proposed here is in general agreement with the one that generally prevails over all other choices.

Table A.1 exhibits a typical cash flow analysis. Note that the current U.S. standard depreciation method, modified accelerated capital recovery system (MACRS), is used but that any scheme can be implemented. This cost justification model allows the investment (expense) to be spread over every year of the planning horizon; tax rate on the nondepreciable portion can be varied as can the portion of the depreciable investment that is tooling. The expected savings (income) can be any rational values. Depreciation rates are fixed by MACRS. Tax rate for income is again variable (if desired). Using zero present worth as the basis, the total investment is readily determined; a portion of that investment is likely to be depreciable.

Let S be the savings in any year, I the investment, $1/\rho$ the depreciable portion, t the tax rate, D_c the capital equipment depreciation, D_t the tooling depreciation, then the net income N in any year is

$$N = (1 - \tau)(S - I) + \tau(D_c + D_t - I/\rho)$$

and the discounted net income for year i (where r is the internal rate-of-return) is

$$\eta_i = \frac{N_i}{(1 + r)^i}$$

MACRS depreciation: capital equipment, 7 years; tooling, 3 years

		Expense forecast				Income forecast		
Year	Ratio	Tax rate	Depreciable	tooling	Savings	Capital equipment	Tooling depreciation	Tax rate
0	100.00%	34.0%	40.0%	0%				
1					1000.00	14.29%	33.33%	34.0%
2					1125.00	24.49%	44.44%	34.0%
3					1265.63	17.49%	14.81%	34.0%
3*					**Salvage value**	43.73%	7.41%	

Total investment = 2768.225; depreciable investment = 1107.290
Capital equipment cost = 1107.279; tooling cost = 0.011
Internal rate-of-return = 14.00%

Projected cash flow						
Year	Income	Capital equipment depreciation	Tool depreciation	Taxes	Net	Discounted net
0	−2768.225	0.000	0.000	−564.718	−2203.507	−2203.507
1	1000.000	158.183	0.004	286.217	713.783	626.126
2	1125.000	271.170	0.005	290.300	834.700	642.274
3	1265.625	193.693	0.002	364.456	901.169	608.263
3*	484.234	0.000	0.000	0.000	484.234	326.844
Income totals	3874.859	623.046	0.010	940.973	2933.886	2203.507
Net totals	1106.634	623.046	0.010	376.255	730.378	0.000

Nominal capital recovery = 1051.580.
Payback in approximately 2.73 years.

Table A.1 Allowable Investment Cash Flow Table—Simple Case

An example with only year zero investment is shown in Table A.1. In Fig. A.1, many of the fundamental criteria are shown. For this example,

planning horizon is 3 years (horizontal axis of graph)

initial investment is the value of both curves at year 0

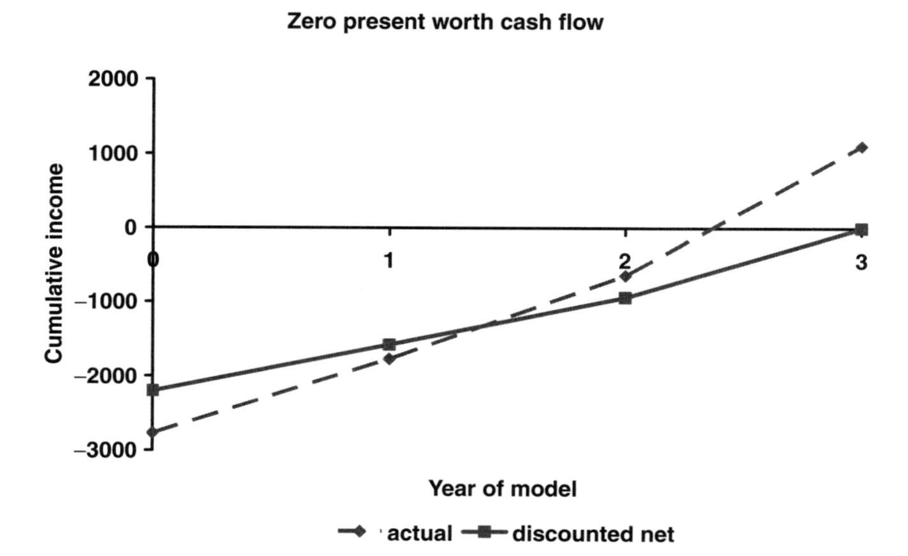

Figure A.1 Plot of income for two simultaneous conditions—simple case.

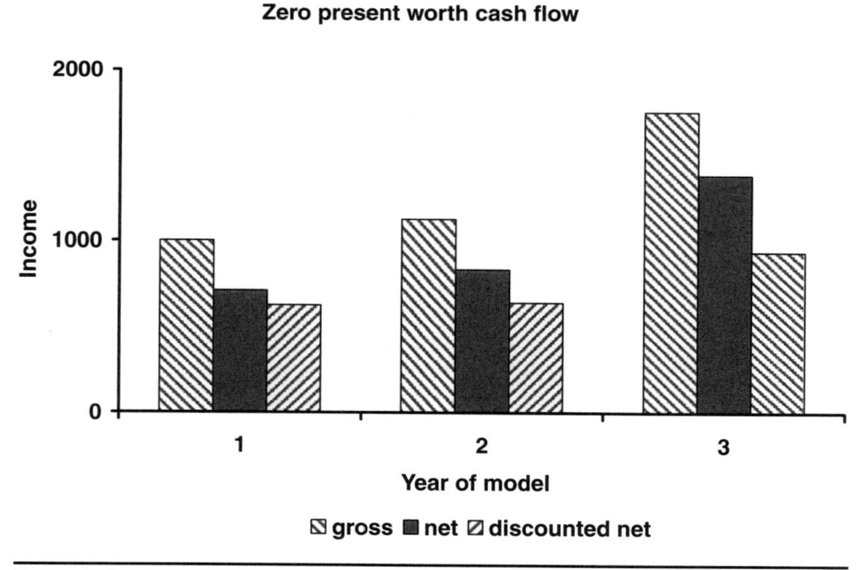

Figure A.2 Bar chart for three simultaneous cash flows—simple case.

MACRS depreciation: capital equipment, 7 years; tooling 3 years
3.5% constant rate of increase in costs

		Expense forecast				Income forecast		
Year	**Ratio**	**Tax rate**	**Depreciable**	**tooling**	**Savings**	**Capital equipment**	**Tooling depreciation**	**Tax rate**
0	74.19%	34.0%	40.0%	35.0%				
1	18.41%	34.0%	80.0%	100.0%	1000.00	14.29%	33.33%	34.0%
2	7.40%	34.0%	80.0%	100.0%	1125.00	24.49%	44.44%	34.0%
3					1265.63	17.49%	14.81%	34.0%
4					1423.83	12.49%	7.41%	34.0%
4*					**Salvage value**	31.24%	0.00%	

Total investment = 3539.541; depreciable investment = 1781.283
Capital equipment cost = 682.727; tooling cost = 1098.556
Internal rate-of-return = 14.00%

Projected cash flow

Year	Income	Capital equipment depreciation	Tool depreciation	Taxes	Net	Discounted net
0	−2625.874	0.000	0.000	−535.678	−2090.196	−2090.196
1	−651.637	0.000	0.000	−44.311	−607.326	−532.742
	1000.000	97.532	122.541	265.175	734.825	644.583
2	−262.030	0.000	0.000	−17.818	−244.212	−187.913
	1125.000	167.199	337.158	211.019	913.981	703.279
3	1265.625	119.427	356.031	268.567	996.968	672.925
4	1423.828	85.305	197.629	387.904	1035.924	613.350
4*	298.462	0.000	0.000	0.000	298.462	176.714
Income totals	5112.915	469.464	1013.358	1132.755	3980.161	2810.851
Net totals	1573.374	469.464	1013.358	534.947	1038.427	0.000

Nominal capital recovery = 1154.139.
Payback in approximately 3.29 years.

TABLE A.2 Allowable Investment Cash Flow Table—Complex Case

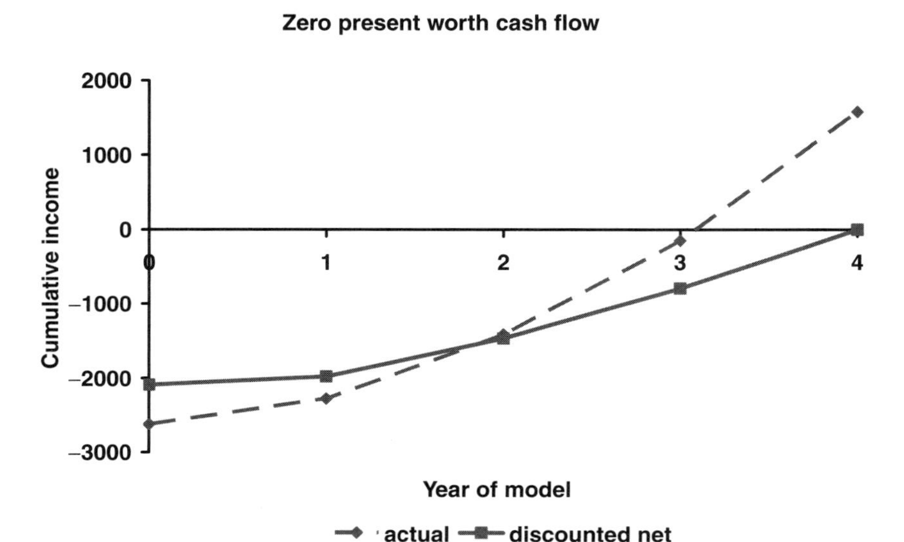

Figure A.3 Plot of income for two simultaneous conditions—complex case.

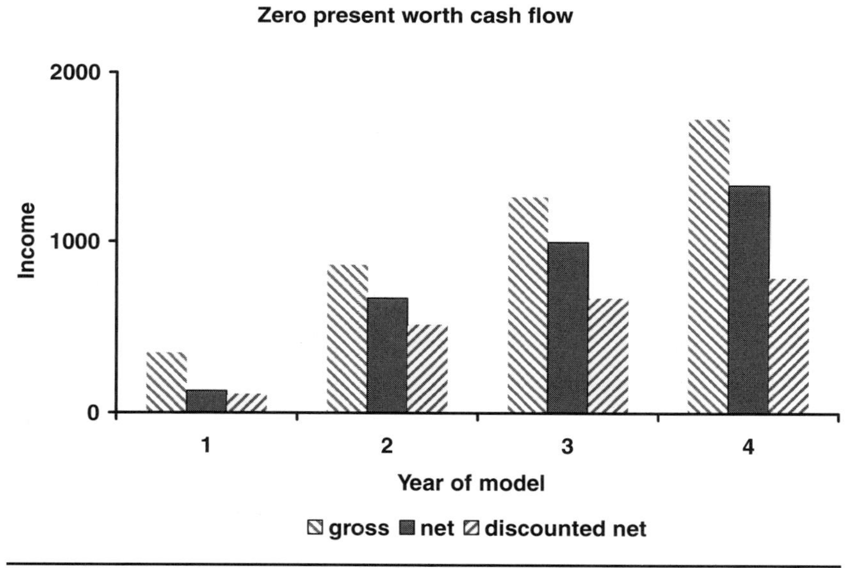

Figure A.4 Bar chart for three simultaneous cash flows—complex case.

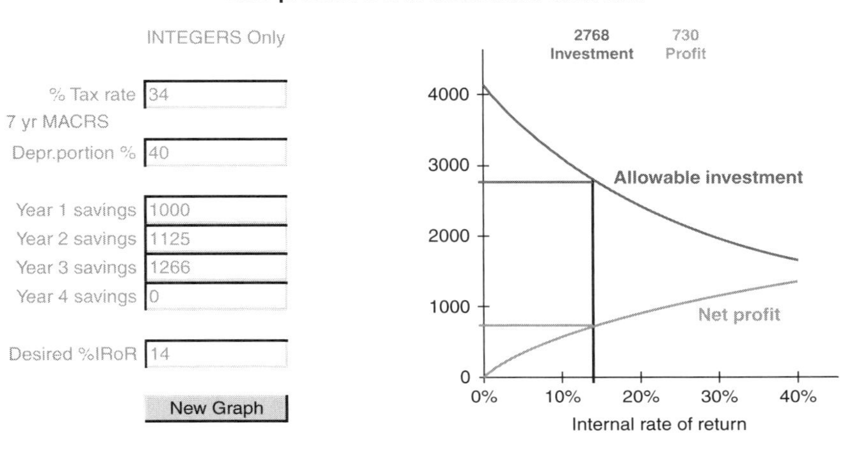

Figure A.5 Interactive web page for a simple case.

payback period occurs when the actual net income curve crosses from negative to positive (in this case, at approximately 2.4 years)

net profit is the value of the actual net income at year 3

Note that the cumulative discounted net income curve ends at zero; this is the definition of **zero present worth** cash flow (i.e., the sum of the future discounted net incomes is equal to the year zero net expense, and therefore there is zero present worth). A bar chart (Fig. A.2) of the income data may also be helpful.

A more complex situation arises when the investment is spread out over a number of years and tooling is separated from capital equipment. For this example, a fourth year of savings has also been added (see Table A.2). Note that the capital equipment and tooling categories are the sum of the various yearly hardware expenditures at the appropriate depreciation values. For this case, Fig. A.3 exhibits the behavior over time. A bar chart of the income (minus expense) data is shown in Fig. A.4.

Examples of the least complex case can be tried online at the web site page www.sysyn.com/intract2.html. This is the most straightforward example possible; there is investment only at time zero, no provision is made for separate tooling depreciation, and the tax rate is constant. For this case,

$$I_{\mathrm{A}} = \frac{(1-\tau)\sum_{i=1}^{H}\frac{S_i}{(1+r)^i}}{\left(1-\tau+\frac{\tau}{\rho}\right)-\frac{\tau}{\rho}\sum_{i=1}^{H}\frac{\delta_i}{(1+r)^i}-\frac{v_H}{(1+r)^H}}$$

A printout of a typical result using the web page is shown in Fig. A.5.

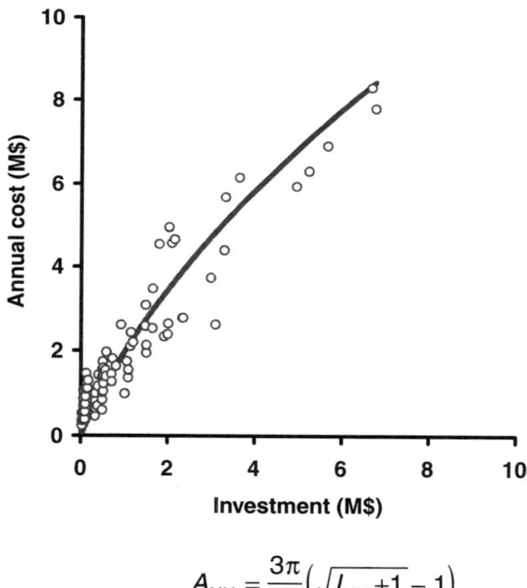

$$A_{\mathrm{MM}} = \frac{3\pi}{2}\left(\sqrt{I_{\mathrm{MM}}+1}-1\right)$$

where A_i = annual cost, million $ (includes labor, oper/main. & capital recovery)

I_i = investment required, million $ (includes hardware, engrg, set-up, etc.)

Figure A.6 Annual cost vs. investment for mostly manual systems.

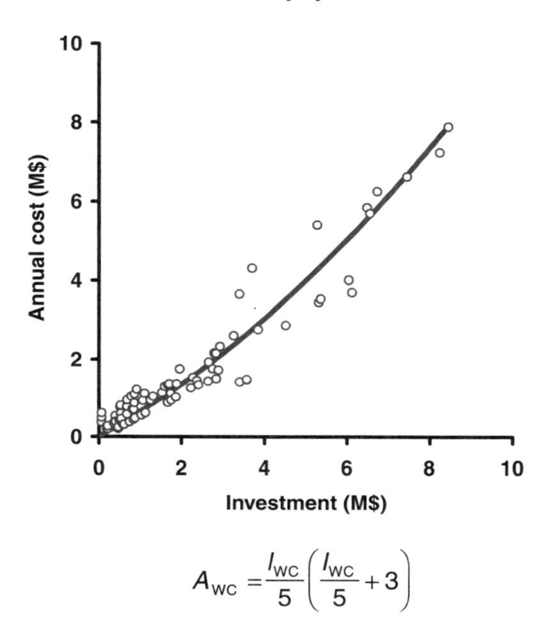

where A_i = annual cost, million \$ (includes labor, oper/main. & capital recovery)

I_i = investment required, million \$ (includes hardware, engrg, set-up, etc.)

Figure A.7 Annual cost vs. investment for world-class systems.

A.3 Allowable World-Class Investment

Based upon a variety of products over a wide range of production requirements, Systems Synthesis, Inc., has established the cost characteristics (using least-squares polynomial fit for **annual cost vs. investment** data and simplifying the coefficients) shown in Figs. A.6 and A.7. Note the difference in curve convexity. Combining the annual cost and investment data for a heterogeneous group of products with a range of production quantity requirements, the characteristic exhibited in Fig. A.8 has been determined. This graph can

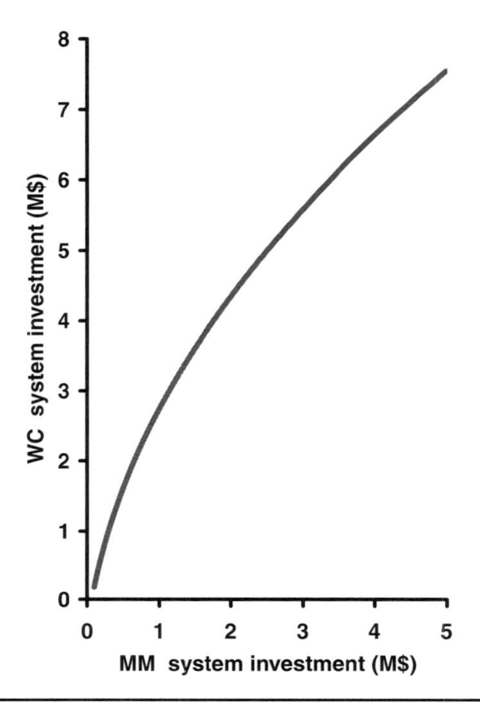

FIGURE A.8 Investment comparison of world-class and mostly manual systems.

be used as a first approximation to the maximum investment for a world-class system when the investment is known or can be specified for a mostly manual current *or* proposed system. In determining this curve, the contribution due to each of three economic parameters: IRoR (internal rate-of-return), years of savings, and ρ (total investment/hardware cost) was investigated. *None* of these factors has any significant effect on the characteristic curve (I_x values in millions):

$$I_{WC} \leq 1.24 \left[-1 + \sqrt{0.3 + 10 \times I_{MM}}\right]$$

This is only a first approximation, and actual system costs MUST be determined!

APPENDIX B

Economic–Technological Synthesis of Systems

B.1 Introduction

The system design method to be described here is a combination of "bottom-up" ideas and "top-down" approaches. Other researchers have attempted portions of such a coupling. Graves and Whitney (1979) and Lamar and Graves (1983) utilized linear programming techniques. Daschenko (1986) used building blocks to optimize process structure variants. Holmes (1987) investigated the shortest path through a network algorithm.

Traditionally, the "bottom-up" approach has been used in manufacturing since no usable technique has yet been widely accepted as "the" way to solve the system design problem. Experienced personnel (industrial and/or manufacturing engineers) use their knowledge of what has worked on similar products in the past to determine a feasible way of making the parts for a product come together. While usable, sometimes after quite a few redesigns, such a system will probably not be the most cost-effective. Nevertheless, such systems are often exhaustively analyzed, particularly using the most elaborate means available—simulation. The latter can take months to bring to fruition and expend significant funds to run the simulation each time. One of the advantages to such an approach is that means for "continual improvement" will usually be available. However, what if the best system could be determined at the beginning of the production run thereby negating, or at least minimizing, the need for such annual betterments?

The goal is to determine a system that has the lowest initial cost and also the minimum possible yearly costs. A rigorous and robust method was created and has been developed over a number of years; it has evolved while helping manufacturers design and install such systems. Thus, it is not just another academic solution useful only for special-case, real-world problems.

B.2 Basic Ideas

The need for a system most often arises when more than four or five fixed sequence tasks must be performed. Ordering of those tasks for the assembly system case is described in App. C; for other situations, there are currently no known automated methods available. In any event, the tasks to be performed need to be specified by some procedure. Each of the tasks must have at least one applicable resource type. Most tasks will have choices among manual, fixed automation, and programmable automation. Each of the resource types applicable to the set of tasks has cost and performance characteristics, which must be enumerated.

Although a simple comparison of the expected costs for a particular resource type to perform each task is possible, it is the combination of tasks, which can be performed by one resource type, that usually provides the best cost-effectiveness. For example, it may be most advantageous to combine tasks 8, 9, and 10 at a workstation of a particular type. This condition can be possible only when there is time available (see Chaps. 3 and 6) at such a workstation.

B.3 Annualized Cost (or Capital Recovery) Factor

At first glance, it may seem difficult to compare the required investment and the cost of manual labor. Actually, it is very straightforward once the investment is annualized. Depending upon who is thinking about the matter, such annualizing is treated as either a capital recovery or an annual cost (they are numerically equal). Generally, an investment is spread out over a number of years H at a prescribed rate of return r and can be thought of as what is currently known as a residual loan. The common equation used is

$$A = I_0 \left[1 - \frac{v_H}{\rho\,(1+r)^H}\right]\left[\frac{r\,(1+r)^H}{(1+r)^H - 1}\right]$$

The residual value at the end of H years v_H is usually taken as the not yet depreciated decimal portion of the part of the initial investment I_0 that can be depreciated as defined by the ρ factor (see App. A). Annual cost factor (or capital recovery factor) is then

$$f_{\mathrm{AC}} = \frac{A}{I_0}$$

B.4 Cost Comparison Equation

There are two primary costs, which are interrelated, to be considered. Unit cost is more often used than the total annual cost (which is the unit cost multiplied by the yearly production volume) simply because the numerical values are easier to grasp and, especially if the product is for consumers, the portion of the selling price due to such manufacturing activity is readily apparent.

System cost is composed of fixed cost (the annualized cost, or capital recovery, of the investment made in the system) and variable cost (labor and operating expenses). As shown in Chap. 3, the fixed cost is a function of the total investment required for the system:

$$C_{\mathrm{F}} = \rho \sum_{n} (P_{\mathrm{R}} + N_{\mathrm{T}} P_{\mathrm{T}})$$

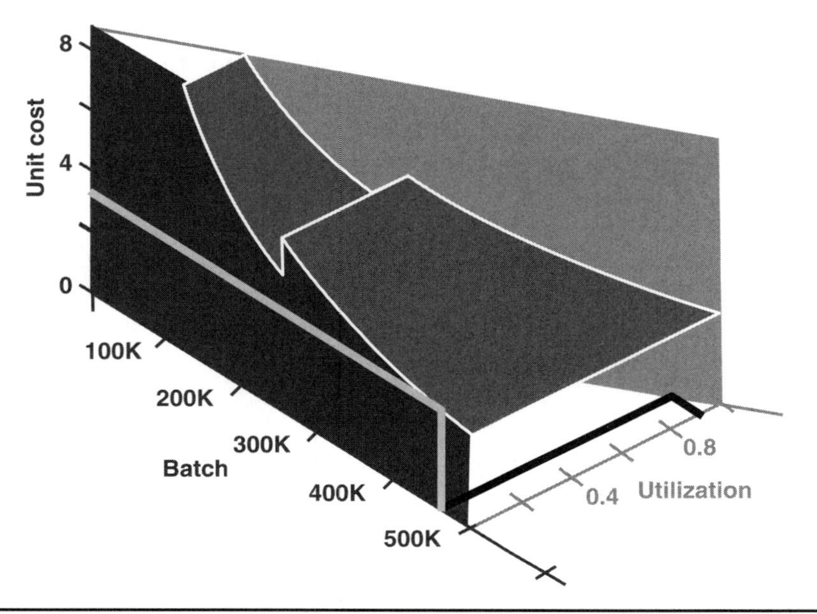

FIGURE B.1 Typical unit cost vs. production volume vs. utilization.

while the variable cost per hour of operation is

$$V_H = O_H + wL_H$$

Unit cost is found by dividing the fixed cost, modified by the annual cost (or capital recovery) factor (see Chap. 3), by annual production and adding the variable rate multiplied by the cycle-time:

$$C_{\mathrm{U}} = \frac{f_{\mathrm{AC}} C_{\mathrm{F}}}{Q_{\mathrm{Y}}} + \frac{V_H T_{\mathrm{CYC}}}{3600}$$

This is the criterion that is to be minimized. The behavior of unit cost vs. yearly production batch is typically like the example in Fig. B.1.* Although the graph was generated for a complete system, each workstation's applicable resource

* The "sawtooth" identifies shift limits—in this example, the maximum for one shift and the minimum for two shifts.

types will have a similar cost vs. production behavior associated with it. What needs to be done is to determine, for a prescribed production batch, the resource type whose behavior exhibits the least cost. It is important to note that the cost equations shown earlier apply to each workstation (in this specific case, $n = 1$). With the constraint that T_{CYC} must be less than or equal to the time available (a function of production volume required), **the goal is to perform as much work as possible for the lowest unit cost at each workstation**. It is thus possible to assure local optimums. While this technique does not absolutely guarantee the global optimum solution to the problem, there is currently no known alternative way to determine anything close to such an optimum for real-world systems.

B.5 Utilization

An important factor, at least for manufacturers, in determining the "goodness" of a system is a characteristic termed utilization. Usually defined as the portion of the available time that is actually required to "get the work done," it is shown as the third axis in Fig. B.1. Utilization (being 0.923 for the example shown) multiplied by the time available identifies the actual cycle-time. Utilization of each workstation may be useful as a goodness criterion when a system is composed of mostly manual workstations and labor is charged for an entire shift regardless of how much work time is actually needed. However, for mixed systems (probably composed of all three generic types of resource types), the best system may require significantly less than the maximum available shift time. This statement is true even when all labor (direct and indirect) is charged for a full shift!

How is such a cost-effective system created?

B.6 Applicable Technology Chart

Although not required for use in the system design procedure, it may be convenient to enter all of the necessary economic and technological data onto a form like that shown in Fig. B.2 (data shown is taken from the App. F example). Note that the task/resource matrix portion, only for an

Title **A.C.M. Final Assy. Demo**

ASDP

APPLICABLE TECHNOLOGY

240 Working days per year
3 Shifts available
5 s Station-to-station move time

14 % Min. Attractive Rate-of-Return
5 yr Capital Recovery Period
19 $/hr Ave. Loaded Labor Rate

RESOURCE data set name ACMRES-D

TASK data set name ACMTSK-D

Production Batch (units/y-)

175000

For each resource type :

C : Hardware cost ($)
ρ : Installed cost / hardware cost - rho factor
e : Up-time expected (%)
v : Operating / maintenance rate ($/hr)
t(c) : Tool change time (sec.)
m(s) : Maximum stations per worker
σ : Tooling cost / Support cost (%)
τ : Maximum tools at a station

When a resource can be used on a task :

Operation time (sec.) : Tool number
Support hardware cost ($)

Number	RESOURCE TYPE / TASK	MNL	PS1	FXD			
		C: 1 ρ: 1.1 e: 87.5 v: .5 t(c): 1 m(s): .83 σ: τ: 6	C: 40000 ρ: 2.5 e: 85 v: .8 t(c): 4 m(s): 5 σ: τ: 4	C: 0 ρ: 1.5 e: 85 v: .8 t(c): 0 m(s): 4 σ: τ: 1	C: ρ: e: v: t(c): m(s): σ: τ:	C: ρ: e: v: t(c): m(s): σ: τ:	C: ρ: e: v: t(c): m(s): σ: τ:
1	Assemble (2), (3), (12), (13), (14), (15) & (16)	26 : 101 1200	24 : 201 20000	11 : 302 170000			
2	Harness (10)	12 : 104 500	15 : 203 2000	X			
3	Evaporator Case (1)	10 : 102 100	8 : 202 10000	5 : 304 60000			
4	Assemble (16), (5) & (6)	32 : 105 3500	27 : 204 13000	X			
5	Automatic Inspection	X	X	14 : 307 100000			
6	Assemble (17), (18) & (19)	36 : 108 7500	31 : 206 15000	15 : 309 150000			
7	Evaporator Core (4) 3 screws	12 : 110 2000	10 : 208 10000	X			
8	Core Shroud (7)-a 2 screws	10 : 110 2000	8 : 208 10000	X			
9	Heater core (7)-b Clamp (20), 3 screws	12 : 110 2000	10 : 208 10000	X			
10	Cover (9) 7 screws	31 : 109 2100	22 : 207 15000	9 : 312 70000			
11	Final Inspection	15 : 113 5000	X	12 : 314 50000			
12	Pack	12 : 114 100	10 : 210 10000	X			
13							
14							

Prepared by ______ Date ______ Sheet 1 of 1

FIGURE B.2 Example applicable technology data.

assembly system, can be generated starting with the methods described in App. C.

At least one resource type must be applicable to each of the specified tasks. Those tasks that are not capable of being performed by a particular resource type are readily discerned. As previously mentioned, the goal is to do as much work as possible within the time constraint at the lowest cost. To accomplish that objective, a combination of tasks at a workstation will often take place. When the production rate is high, there will be little time available and thus combinations of tasks cannot take place unless replicated stations doing more than one task turn out to be cost-effective; this situation rarely happens for a one-product system but often occurs for multiple product systems.

B.7 Finding the Least-Cost System

The scheme used to find least cost is quite straightforward. Essentially, the expected unit cost for various combinations is determined and the lowest one selected. For example, the scenario for the demonstration input data would be the following:

Product 1: [A CM Final Assembly DEMO]					
	Task				
Resource	**01**	**02**	**03**	**04**	**05**
MNL	0.7603	0.3816	0.2550	0.1940	
PS1	0.3708	0.1918	0.1422	0.1216	
FXD	0.5522				

There is time to perform up to four tasks at this station for two of the available resource types. In this case, tasks 1, 2, 3, and 4 would be assigned to a PS1 resource type. Note that the entire resource cost as well as all labor cost will be allocated at the first task assigned. As a check on the method, let's determine the unit cost for this workstation. Recall that the resource and support costs shown above are annualized.

Therefore, the station fixed unit cost is

$$C_F = \frac{27778 + 13889 + 1389 + 6944 + 9028}{175000} = 0.3373$$

while the station variable unit cost is

$$C_V = \frac{23215 + 869 + 549 + 1418}{175000} = 0.1489$$

Adding these two costs produces 0.4862 as the unit cost to perform four tasks. Thus, the cost per task at this workstation is expected to be

$$\text{Cost per task} = \frac{0.3373 + 0.1489}{4} = 0.1216$$

The principal cost advantage here is that only one resource type is required for the four tasks. In a similar manner, the cost per task for this resource type to perform 1, 2, or 3 tasks would be

$$C_1 = \frac{27778 + 23215 + 13889}{(1) \times 175000} = 0.3708$$

$$C_2 = \frac{27778 + 23215 + 869 + 13889 + 1389}{(2) \times 175000} = 0.1918$$

$$C_3 = \frac{27778 + 23215 + 869 + 549 + 13889 + 1389 + 6944}{(3) \times 175000}$$
$$= 0.1422$$

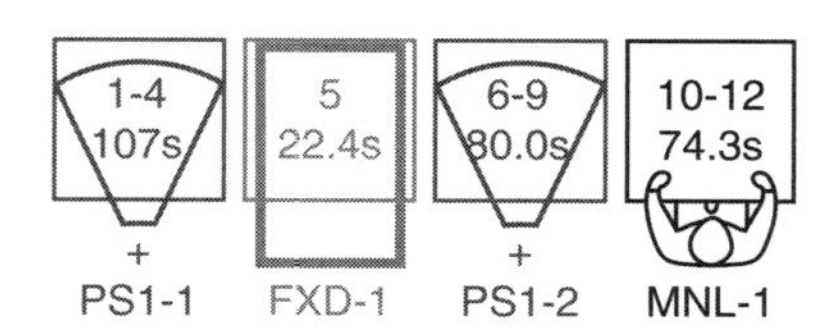

FIGURE B.3 System schematic for example.

Resource	Total cost	Number used	Time used	Unit Cost		Number of		
				Fixed	Variable	Tasks	Tools	Workers
MNL	135725	1	74.3	0.012	0.764	3	3	1.20
PS1	155218	2	107.1	0.595	0.292	8	6	0.40
FXD	68546	1	22.4	0.230	0.161	1	1	0.25
Total				**0.837**	**1.217**			**1.85**

Minimum pallets in system = 3.
33.63 units per hour.
107.1-s cycle-time expected.
85.00% bottleneck station uptime expected.
2.90 $/h system operating/maintenance rate.
359489 cost ($) to produce 175000 units, with unit cost ($) = 2.054.
System annualized charge factor = 0.2750.
532921 ($) total investment required.
257201 ($) for required hardware.

Table B.1 Demonstration System Cost and Performance Results

All anticipated unit costs are determined in the same manner. The situation is more complex for multiple products but proceeds along essentially the same lines. The next workstation must begin at task 5. In this case, task 5 would be assigned to the only applicable resource type (FXD). The next workstation must begin at task 6. Subsequent resource allocation uses the same procedure. The behavior of the resulting system includes the schematic shown in Fig. B.3 and the characteristics of the system are seen in Table B.1.

Every system can be synthesized in the same manner. When multiple "products" are to be produced by a single system, the costs for each are determined as shown in Chap. 6. The expected cost of the combination is then ascertained. Thus, any system wherein a specified set of tasks with a prescribed set of possible applicable resource types has been defined can be least-cost synthesized.

APPENDIX C

Establishing Task Data for Assembly Systems

C.1 Introduction

Significant work has been done on the creation of computer models for products or sections thereof. Surface modeling, solids modeling, variational geometry, parametric design, and cutter path determination have all been important additions to improving product design and the fabrication of the appropriate components. Design for assembly principles is now generally applied and will reduce product expenses as long as the remaining parts do not produce an *increased* aggregate cost. The method to be discussed here presumes that all of those techniques have been properly implemented before it is started.

Almost every product is composed of more than one component; every production company must therefore face the prospect of actually putting those parts together. Information about the product must be transformed into a workable assembly process plan that should form the basis for designing the best assembly system (Gustavson, 1990). A primary assumption here is that the breakdown of a product into subassemblies has already occurred on a functional basis and has resulted in nonseparable, easily tested units. The technique to be described deals with only *one* such assembly or subassembly at a time. At any level of assembly, there generally will be less than 25 specifiable components (which may themselves be subassemblies).

When product designers investigate assembly using current software, little beyond geometric interference can be checked. Significantly more than that often needs to be done if the actual assembly is to be as cost-effective as possible. What has been needed is a method that creates the best available assembly process plan; this appendix describes the only known automated technique. Fundamental elements of the method are

create/extract product data
develop/alter an exploded view
establish/change an assembly sequence
specify intermediate test requirements
determine/alter an assembly process plan

Starting with the sample product sketch in Fig. C.1, the corresponding part and mate graph shown in Fig. 3.1 can be created. Such a graph can be very useful for understanding how a product must be assembled.

When all phases of the method have been implemented in software, you can be involved at every step of the process. Important insight into the world of assembly systems and how product design decisions affect their behavior and cost can significantly benefit product designers and engineers.

Because the technique can be rapidly utilized, it is valuable at any stage of product/process design. Fundamental ideas are discussed in Gustavson (1990), and the techniques have been importantly expanded and improved. Background material discussing the steps involved and showing the need for automatically bridging the gap between products and assembly systems can be found in Nevins and Whitney (1989). The method discussed in this appendix has been successfully applied to a variety of products.

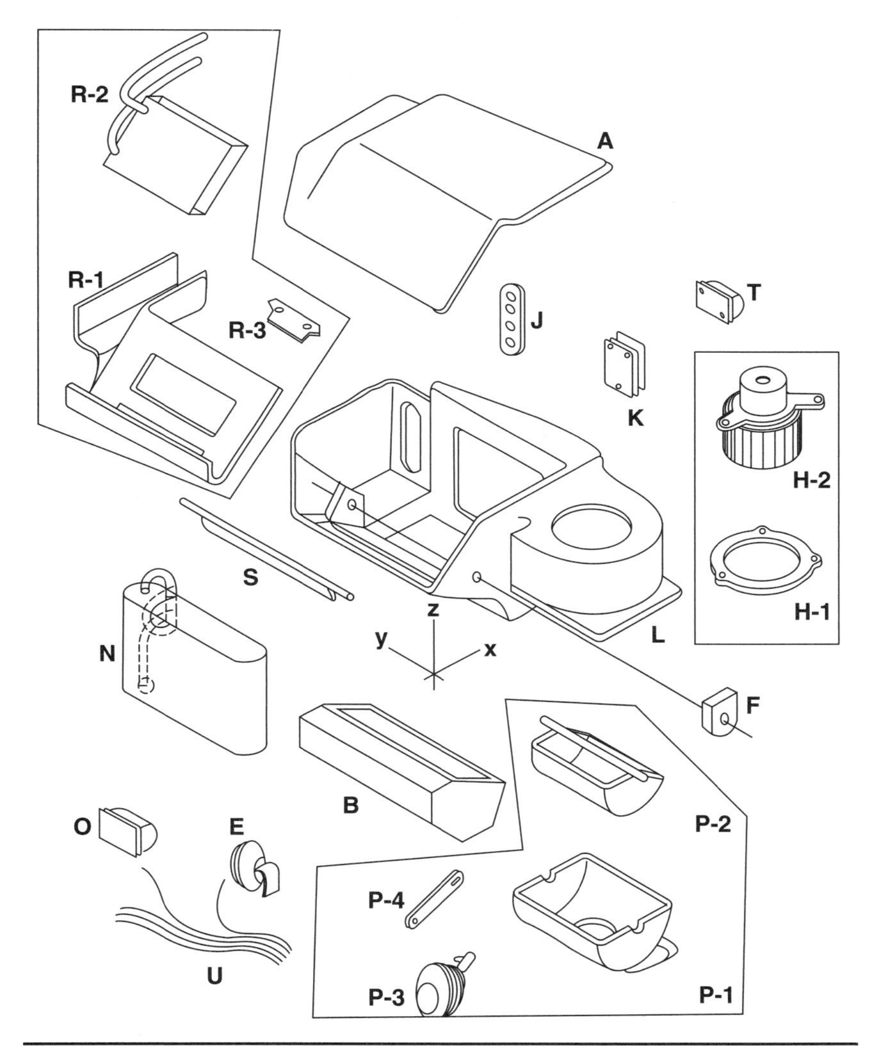

Figure C.1 Sketch of a sample product.

C.2 Fundamental Principles

Every assembly system must have a coordinate frame specified; it may have little relationship to the one(s) used to design the assembly. The axes are to be oriented such that the principal assembly direction is negative *z* (most components will be assembled downward thereby making best use of gravity).

Each component (single part, subassembly, or collection of parts that must be assembled as a group) is specified by a code letter, the weight, a bounding box, and a description. Some of these data are derivable from existing CAD information, although none has been officially integrated.

Every component must have at least one *physical* mate with another component in the assembly. Many of them will have a group of physical mates. Fourteen generic mate types are currently implemented: some have fit requirements and/or multiple action needs.

The integration of component and mate data produces an exploded view of the assembly, an assembly sequence, and the best assembly process plan. Every step allows manual editing.

C.3 Input Data Requirements

Component Data: Geometric data is defined in the totally assembled location by using the assembly coordinate system (principal assembly direction should be negative z).

1. Code—an arbitrary, single character (usually a letter) used to keep track of components.
2. Weight—used to help determine size and cost of resource types.
3. Bounding box dimensions—the planes defining minimum and maximum dimensions in x, y, and z. This data also defines the pictorial representation of the components in the exploded view; some will be right-circular cylinders rather than rectangular boxes.
4. Description—if more than one item is required, separate with semicolons.

This component data can be entered in any convenient order; no pattern is required. It can be readily edited.

Mate Data: *Physical* relationship between any two components. A component may have numerous mates, but must have at least one. Heuristic data indicates that you can expect to find

$$\text{number of mates} = 1.1 \times (\text{number of components})^{1.1}$$

but the value for any particular assembly may differ significantly from this estimate. In any convenient order, the specifications for each mate will be

component codes—the two components that possess the relationship

type of mate—selected from

1. adhesive bond—in general, a three-step process: spread adhesive, mate parts, and oven cure; the first and last steps may be subsequently deleted
2. bearing race/bushing—the peg/hole condition, requires a fit category
3. bolted joint (without spec.)—parts are held together with threaded fastener(s) that do not require a tightening torque specification
4. critical alignment—particular care must be exercised during placement
5. cocure—used for composites in various processes
6. gear(s)—requires a fit category, may be multiple
7. general placement—parts touch but have no particular relationship (e.g., washer/screw)
8. rivet(s)—requires a fit category, probably multiple
9. selective fit—one of the components must be chosen from an available group of sizes
10. snap fit—force may be required
11. solder—usually requires fixturing
12. spline(s)—requires a fit category, may be multiple
13. torqued fastener(s)—generally a three-step process: align parts, insert fastener(s), torque fastener(s); may require a fit category
14. ultrasonic weld—usually requires fixturing

Some task types require additional data:

number required—simultaneous events that must take place; normally 1, unless used to define multiples (e.g., screws)

fit category (simplification of ISO Preferred Fits and ABC Standards):

1. running
2. clearance or drive
3. transition or press
4. interference or shrink

direction—mate axis within a 45° cone (x, y, or z).

Principal assembly direction is negative z.

Each of these mate types has a degree-of-difficulty associated with it; this value may significantly affect subsequent portions of the assembly procedure. Details of the mate data can be altered before proceeding to the next step.

C.4 Exploded View of the Assembly

A pictorial of the assembly that exhibits the general order in which components are added to the base component will provide a significant first step to establishing a usable assembly sequence.

The assembly picture to be produced is composed of rectangular boxes and possibly some cylinders. This representation will generally be adequate since the specific geometric details of the components (defined before starting this procedure) are not usually required.

Each component will have an assembly axis ($-x$, $+x$, $-y$, $+y$, $-z$, $+z$); note that the actual assembly direction is not required to be along that axis, but will normally occur within a 45° cone about any parallel axis. The exploded view will exhibit each component's bounding box and code in the appropriate assembly order along its axis.

C.5 The Base Component

Although the initially assembled component (usually called the base) is generally the largest, that will not always be the case. This procedure establishes the base by evaluating the following for each component:

1. weight—input data
2. enclosed volume—determined from bounding box data
3. sum of degree-of-difficulty of its mate(s)—this factor has a higher weight in the decision

It may also be important, in a particular application, to know whether the base is at the top, at the bottom, or in the middle. Although seldom necessary, automatic selection of the base component and/or its location can be altered.

C.6 The Exploded View

The relative location of all components in an assembly is often shown in an exploded view. Such a pictorial helps to visualize how the components need to be put together. A technique for determining the order of the components along the z-axis (defined to be the principal axis) is a multistep process. Once the base is ascertained, the relative location of all other components is initially found from the following:

1. Use mate direction(s) to define axis orientation
2. Determine location of geometric center relative to base geometric center in order to establish whether the part is on the negative or positive side of the base (in the x, y, or z direction).
3. Position along the axis is determined by the distance to the "away" plane (i.e., the bounding plane farthest away from the base geometric center in the assembly direction).

The second arrangement of principal axis components utilizes the physical mate information to rearrange the order along negative z and positive z. Requirements for complex assemblies may produce significant changes. A third reordering, using the same basis, also takes place automatically. During the development of this technique, 10 such reorderings

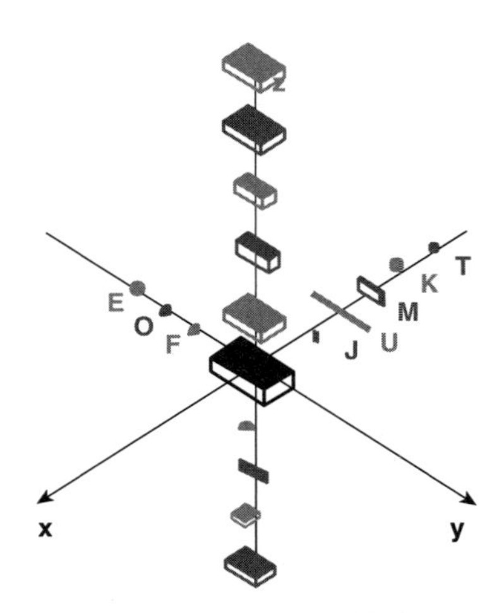

(A) Cover
(G) Cover-to-case sealant
(R) Heater core; heater core
(N) Evaporator core subassy.
(H) Motor & fan; isolator
(L) Evaporator case
(C) Motor-to-case sealant
(S) Temperature valve
(P) Al case; Al valve; vacuum
(B) MVH subassy.

Figure C.2 Initial exploded view of sample product.

were tried; changes rarely occurred after the third pass. It may still be necessary to refine the orientation order for a complex product.

An example exploded (bounding box) view is shown in Fig. C.2. This initial exploded view is quite close to the desired final version. Although such an intermediate result is highly desirable, it will not necessarily occur. The assembly shown needs only one correction from the original display; component L is logically placed between H and C initially. Simply moving L between S and P provides the view in Fig. C.3. Moving any part along its orientation axis, but only between two currently adjacent components, is easy. Although the x-axis and y-axis components can also be moved, they seldom need to be since they are commonly related to z-axis components but not to each other. Once the exploded view is satisfactory, an initial assembly sequence can be determined.

C.7 An Assembly Sequence

Starting with the base component, an order in which all other parts are to be assembled is determined. The general procedure is as follows:

1. Establish the initial group of parts to be assembled (i.e., determine the initial assembly axis) by checking

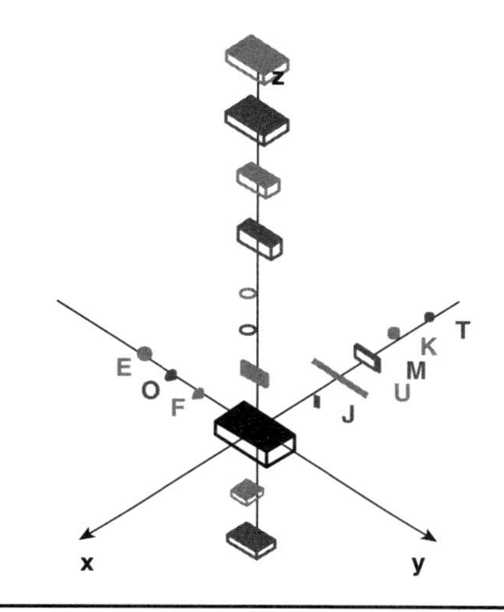

(A) Cover

(G) Cover-to-case sealant

(R) Heater core; heater core

(N) Evaporator core subassy.

(H) Motor & fan; isolator

(C) Motor-to-case sealant

(S) Temperature valve

(L) Evaporator case

(P) Al case; Al valve; vacuum

(B) MVH subassy.

Figure C.3 Revised exploded view of sample product.

the number of components and the difficulty of the mates.

2. Ordering of the groups of components is usually: bottom, left, back, right, front, top.
3. Only a few of the assembly directions may actually contain parts. The customary order can be automatically overridden to satisfy particular requirements of an assembly.
4. For each assembly direction, the following needs to be done:
   a. Set up the exploded view order as the initial sequence.
   b. Using that sequence, check the mate conditions between adjacent parts. Resequencing will automatically occur if undesired behavior is detected.
   c. When investigating the principal axis, check to see whether perpendicular plane components are related. If so, insert them into the sequence.

Any assembly sequence can be altered by merely inserting task *n* in front of any other task *m*. The final sequence derived from the exploded view has had one alteration from

Current assembly sequence

1 (P) Al case; Al valve; V	10 (U) Harness
2 (B) MVH subassy.	11 (M) 3 Stud seals
3 (L) Evaporator case	12 (K) Resistor assy.
4 (S) Temperature valve	13 (T) Solenoid assy.
5 (F) Temperature valve Ac	14 (N) Evaporator core sub-
6 (O) Solenoid #1	15 (R) Heater core; heater
7 (E) Vacuum element #2	16 (J) Pipe seal
8 (C) Motor-to-case sealan	17 (G) Cover-to-case sealan
9 (H) Motor & fan; isolat	18 (A) Cover

Figure C.4 Final assembly sequence for sample product.

that found automatically; L originally preceded P. While it was a logical conclusion based upon the exploded view and input data, the initial sequence required an x–z plane rotation that was cleverly eliminated by a manufacturing engineer who saw that P and B could be placed into the assembly system pallet and L could be snap-fit to both simultaneously. The final sequence is shown in Fig. C.4.

C.8 In-Process Testing

Once the sequence is satisfactory, the potential need for in-process tests must be addressed. To reject assemblies at the least value-added station, intermediate tests can be specified. There is not necessarily a need for such in-process testing. A final test is automatically added, but can be removed from the assembly process plan if not considered necessary.

C.9 The Best Assembly Process Plan

Orientation of the product during assembly can have significant effects on the process plan. There are two choices:

1. Downward assembly only—the assembly must be reoriented so that the addition of every part must be essentially straight down. This has been found to be a basic requirement for high precision products.

2. Fixed orientation—the assembly will generally not be reoriented, although a 180° change in assembly direction from task $n - 1$ to task n will automatically insert a pallet rotate step. It is assumed that all perpendicular direction parts can be added without significant reorienting.

The task description data, which constitutes the assembly process plan, will be stored in a specific file and is therefore given a unique name. It may be found desirable to compare two, or more, plans for the same assembly.

Addition of each task to the process plan utilizes the following:

1. Check to see whether the part has a mate with any component already in the assembly. If not, add a fixture (which must be removed later in the process plan).
2. When multiple mates are possible, it is assumed that they can only occur sequentially. This condition generally raises the task degree-of-difficulty.
3. If reorientation is necessary, add the required step(s).
4. For the mate types that require multiple steps (e.g., adhesive bond), delineate the steps.
5. Add the in-process test if one has been specified to follow addition of a component.

The last two steps in the process plan are automatically defined to be final test and pack/unload assembly for shipment (which might be input to the next level of assembly). Final test can be deleted if it is not required.

At this point, information necessary to display the initial assembly view is available. The data shown in Fig. C.5 ensues. There are two types of data displayed: component tasks and additional tasks. Each component task can be color coded according to

blue—straightforward task

green—some difficulty likely

yellow—hard to do; fixed automation not possible

red—arduous; must be done manually

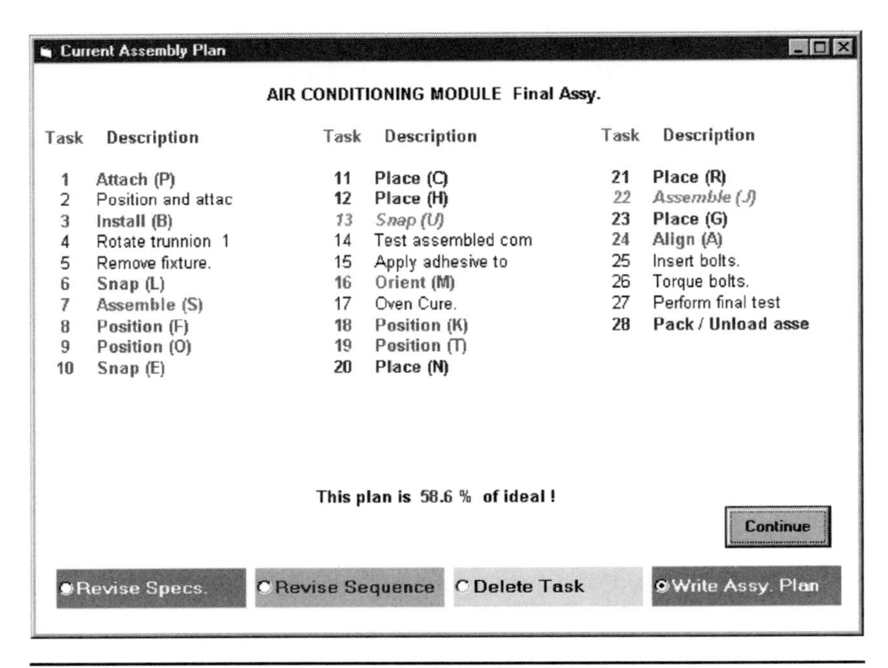

Figure C.5 Initial assembly process for sample product.

All noncomponent tasks can be shown in gray. Only a noncomponent task can be deleted.

Note that this plan is 58.6% of ideal (the SCORE is 58.6); the maximum value of 100% would occur if all tasks but one were straightforward, and the only noncomponent task is pack/unload. There are three ways to alter this data:

1. Delete a task—see the next section.
2. Revise the sequence—this *may* produce a better SCORE by changing the degree-of-difficulty for some of the tasks.
3. Revise the specifications—changing the component and/or mate data *should* provide the most dramatic change in the tasks and therefore in the SCORE, but the latter will not necessarily improve.

By eliminating the tasks that are unnecessary, the SCORE may be significantly increased. Deletions that can be made for this example are as follows:

Task 2—fixture is not needed since the pallet will be designed to hold both (P) and (B).

Task 4—pallet rotate is an artifact of shifting the base component from the first task.

Task 5—fixture was not required; it therefore does not need to be removed.

Task 15—applying adhesive not required since M contains self-adhesive.

Task 17—oven cure not required for this adhesive application.

Task 25—although a tightening torque has been specified, these screws are being driven directly into the FRP case (L); a special inserting station will not be required.

Eliminating all of those unnecessary tasks will definitely change the SCORE, usually for the better. Fig. C.6 exhibits the final ASSEMBLY PLAN. All unnecessary tasks have been

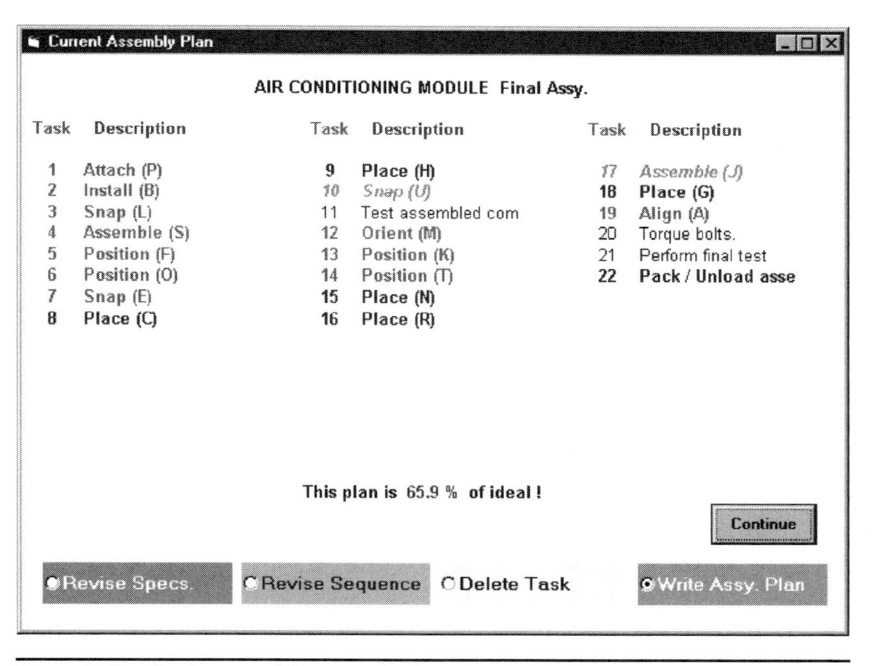

Figure C.6 Final assembly process for sample product.

deleted, thus raising the SCORE to 65.9% of ideal. Note that there are still two arduous tasks; it is generally desirable to reduce the difficulty of such tasks but here it is not possible. One task (10) requires attaching a wiring harness in multiple-planes, while the other (17) requires pushing a seal over four pipes simultaneously and then aligning it with the case (L)

Automatic assembly planning also includes a determination of two characteristics that are important in determining the parameters for resource types applicable to each task that must be performed:

1. Envelope size—simply the difference between maximum and minimum dimensions (checking all the components) for the *x*-, *y*-, and *z*-axis.
2. Relative assembly difficulty—this is a ***critical*** system design parameter since it is used as the power to which time and cost calculations are raised. The numerical value is

   relative assembly difficulty
   = [(total for tasks degree-of-difficulty)/
   number of tasks]$^{0.25}$

Fig. C.7 shows the assembly process plan for this example. The headings are as follows:

Task—the final sequential numbering

Type—an internally-used code that generically defines the task type

Motions Required— X, Y, Z, X–Y, X–Z, Y–Z, or X–Y–Z

Load—usually the weight, but could be the force or the torque equivalent

Dgree of Diffclty—task can be color-coded; see the description above

Task Actns—required multiple; number of components in set, or fasteners required

Task Description—what must be accomplished

Assembly Plan

Task	Type	Motions Required	Load	Dgree of Diffclty	Task Actns	Task Description
1	A	Z	4.00	2	1	Attach Al Case; Al Valve; V
2	I	Z	4.00	2	1	Install MVH Sub-Assy. (B).
3	E	Z	5.00	2	1	Snap fit Evaporator Case (L
4	A	Z	1.00	3	1	Assemble Temperature Valve
5	T	Y	2.00	2	1	Position Temperature Valve
6	T	Y	1.00	2	1	Position Solenoid #1 (O) a
7	E	Y	1.00	2	1	Snap fit Vacuum Element #2
8	P	Z	0.10	1	1	Place Motor-to-Case Sealant
9	P	Z	4.00	1	2	Place Motor & Fan; Isolato
10	E	X	1.00	4	1	Snap fit Harness (U) into a
11	M	X	0.00	2	1	Test assembled components.
12	I	X	0.20	3	1	Orient 3 Stud Seals (M).
13	T	X	1.00	2	1	Position Resistor Assy. (K)
14	T	X	1.00	2	1	Position Solenoid Assy. (T)
15	P	Z	4.00	1	1	Place Evaporator Core Sub-A
16	P	Z	7.00	1	3	Place Heater Core; Heater C
17	A	X	0.50	4	1	Assemble Pipe Seal (J).
18	P	Z	0.20	1	1	Place Cover-to-Case Sealant
19	A	Z	1.00	2	1	Align Cover (A).
20	B	Z	14	2	7	Torque bolts.
21	V	XYZ	0.00	2	1	Perform final test.
22	P	XYZ	38	1	1	Pack / Unload assembly.

Continue

FIGURE C.7 Final assembly process plan for sample product.

With this best assembly process plan (such as shown in Fig. C.7 for the example product), the assembly system design procedure (see Chap. 3) can be started.

C.10 Summary

Starting with basic data about the components and how they physically relate to each other, it is now possible to rapidly define an assembly sequence and proceed to establishing the best assembly process plan. All of this can be done totally automatically, but you can provide knowledge and insight to possibly improve the results at every step of the process.

APPENDIX D

Simultaneous Improvement in Yield and Cycle-Time

D.1 Introduction

The traditional method for trying to understand production behavior during the start-up phase, or when total production is limited, is usually based upon "learning curves" (Hodson, 1992). While many interpretations of the data have been made, the fundamental condition is usually defined by

$$y = y_0 x^{-n}$$

which means that for the xth unit, the value y (usually time, but sometimes cost, etc.) will be some portion of the initial value y_0 as calculated using the "learning rate" n, expressed as a decimal. Another interpretation of this equation states that as the number of cycles x doubles, y decreases by a fixed percentage $n \times 100\%$. The data is often exhibited on a log–log graph such as shown in Fig. D.1. Fig. D.2 exhibits

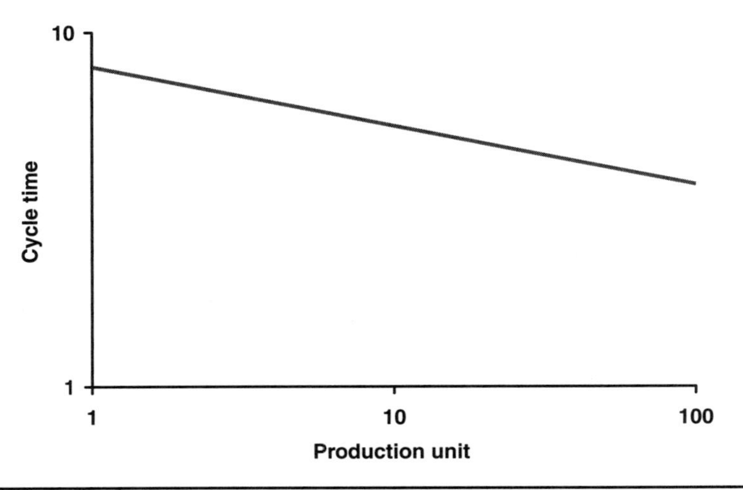

Figure D.1 Cycle-time vs. unit produced—logarithmic.

the same data when plotted using arithmetic scale axes. The latter curve illustrates that learning for early units is much more rapid than that which occurs for later units. Note that there is no "concluding" condition; in theory, the improvement behavior could continue indefinitely.

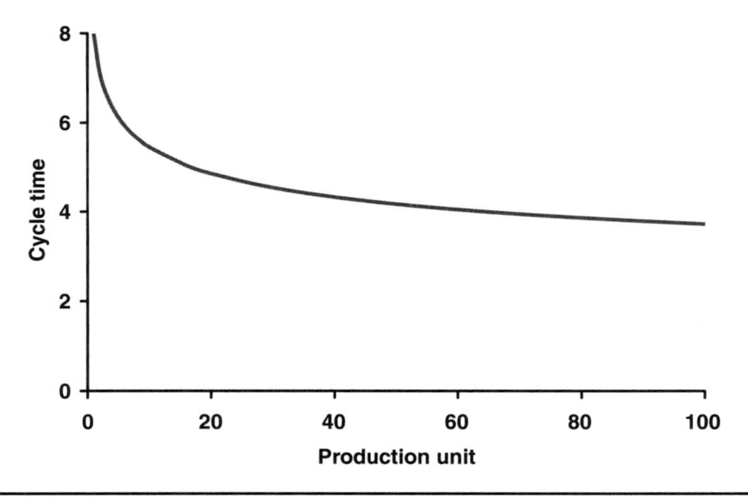

Figure D.2 Cycle-time vs. unit produced—linear.

D.2 A Different Approach

Although the method described above has gained wide acceptance, it has distinct limitations. The primary shortcoming involves the fact that only one operating characteristic (usually cycle-time, but possibly process yield or cost) can be investigated at a time. Also, the traditional method never reaches a "steady-state" condition. The technique to be described here will provide a mathematical basis for simultaneous reduction in cycle-time and improvement in yield; each characteristic will also finish the "learning" phase with constant (i.e., unchanging) behavior.

When plotted using actual (not logarithmic) values, each attribute will be defined to be a portion of a common ellipse of the type:

$$\frac{(x - x_0)^2}{a^2} + \frac{(y - y_0)^2}{b^2} = 1$$

Looking at cycle-time, which is being reduced for each unit processed, the behavior is (data used is from a particular example) shown in Fig. D.3. For this example, the TIF (time improvement factor) is 84.43. Note that the shape is similar

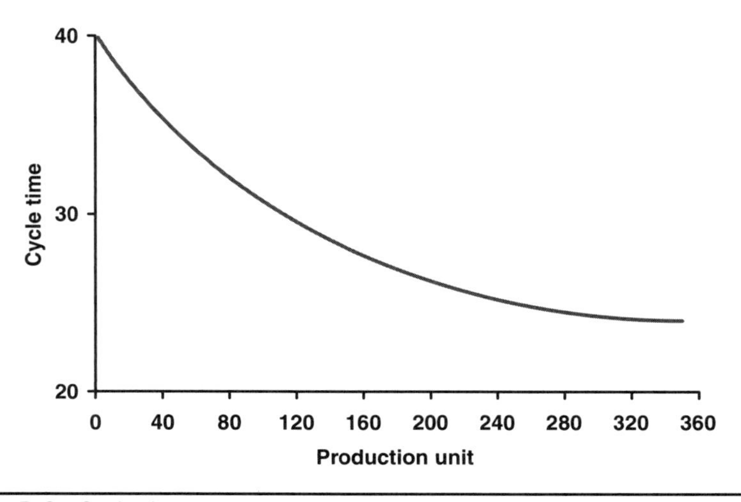

FIGURE D.3 Cycle-time improvement behavior.

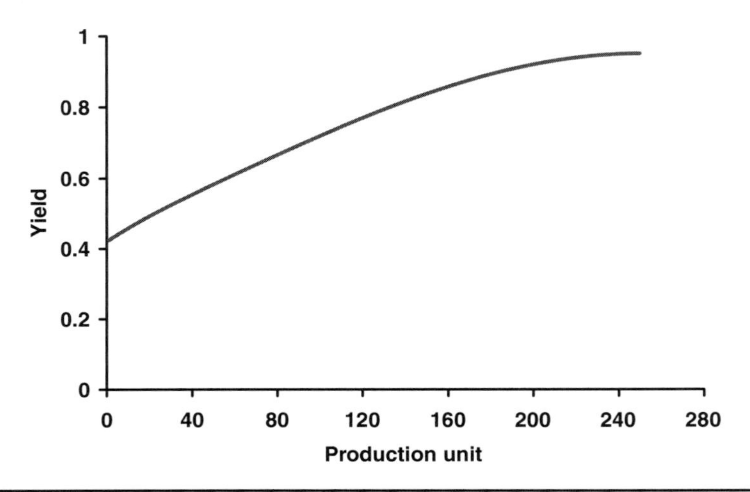

Figure D.4 Yield improvement behavior.

to that for the "traditional" logarithmic curve shown in Fig. D.2. However, the final section of the new curve is horizontal (constant). Time reduction applies to every unit processed during the learning phase.

For yield, which is being increased as each successful unit is produced, Fig. D.4 exhibits the behavior (data is from the same example) that will actually be evaluated using the inverse (or, $1/y$); the resulting curve is shown in Fig. D.5. For this example, the YIF (yield improvement factor) is 77.44. Here, the behavior is similar to cycle-time reduction but is applied to "good" units only. When cycle-time reduction and yield improvement occur simultaneously, the combination is termed PIB (production improvement behavior). For the example above, the CIF (combined improvement factor) is 80.86.

D.3 Evaluating Production Improvement

Solution of the problem requires knowledge of

1. total time available (expressed in days, hours, minutes, *or* seconds)

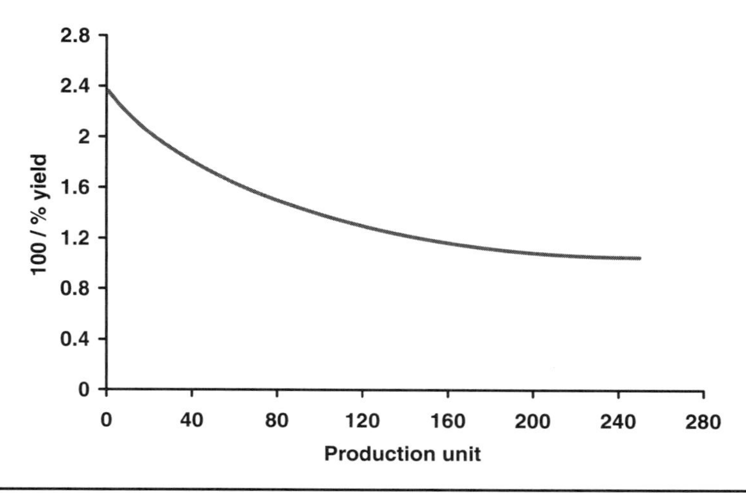

Figure D.5 Inverted yield improvement behavior.

2. expected cycle-time for the first unit (using the same time units)
3. expected yield for the first unit (in percent)
4. functional production units required (either total or "ramp-up")

Knowing the total time available for production along with the working days per year and number of shifts provides the basis for determining how many years, or portions thereof, will be required to complete production or the learning phase of an intended long production run. Specification of the initial yield and cycle-time establishes the limits on the behavior of the two parameters that are to be improved.

As an example, utilize portions of the data shown in Fig. D.6. For 10,000 total hours, a period of

$$\frac{10000}{240 \times 2 \times 8} = 2.6 \text{ years}$$

will be available to process the required units.

Specification of initial yield and cycle-time result in an allowable range of 83 to 228 output units. Fig. D.7 exhibits

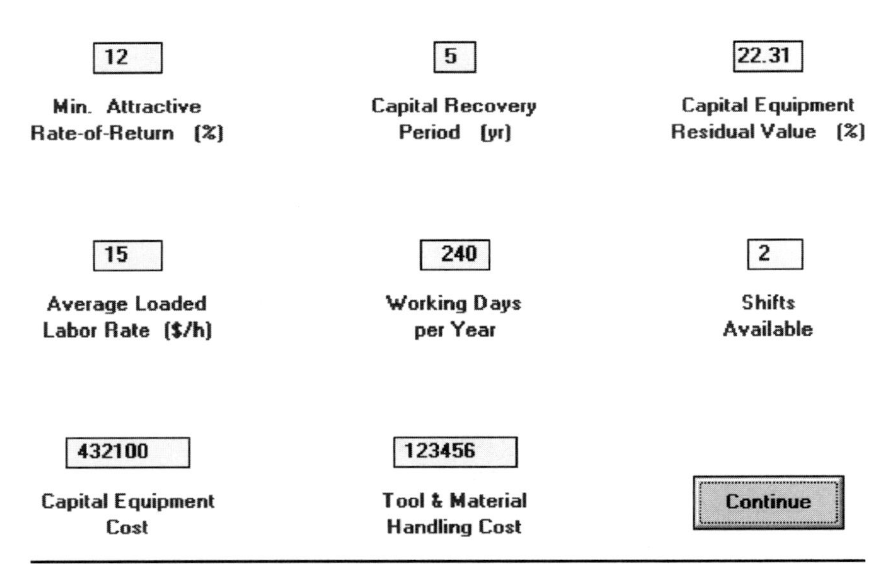

FIGURE D.6 Typical economic and time data.

the general behavior; data similar to that shown can be determined using the interactive web page www.sysyn.com/pibactv2.html. Since the yield will start at less than 100%, the units needing to be processed will be greater than the units output, sometimes considerably so.

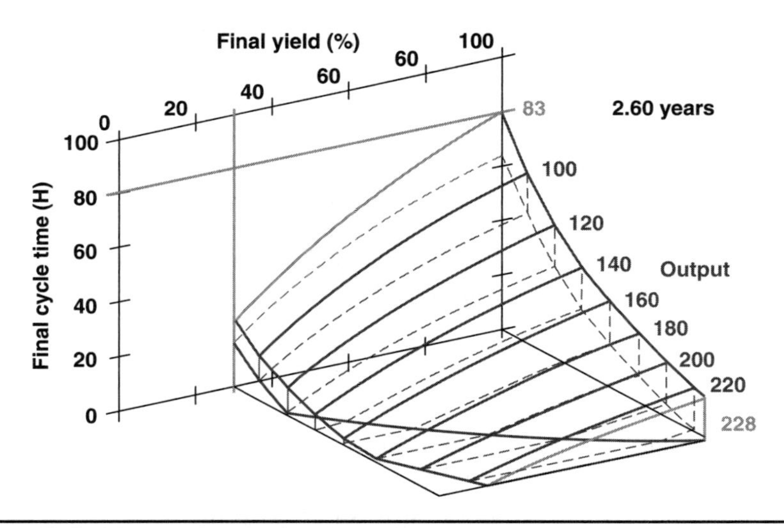

FIGURE D.7 Yield and cycle-time behavior—general case.

After selecting the required output units, the desired characteristics for the final unit must be specified:

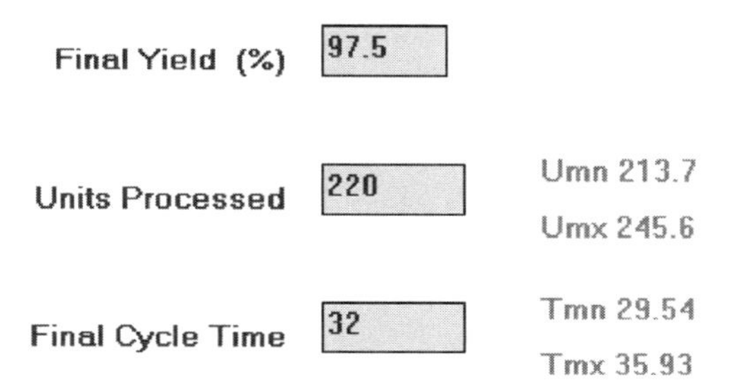

Note that the acceptable range for units processed and for final cycle-time (both of which must be calculated) can be established as shown above. Particular values within each range are readily specified.

A plot (Fig. D.8) of final cycle-time vs. final yield for a specific total output will exhibit an uninterrupted area wherein all usable values of both characteristics (cycle-time and yield) must lie. The desired final values are denoted by a "dot" within that area. Any combination of final yield and final cycle-time that lies within the solid area shown in Fig. D.8 could be selected. For every such choice, a YIF and TIF can be

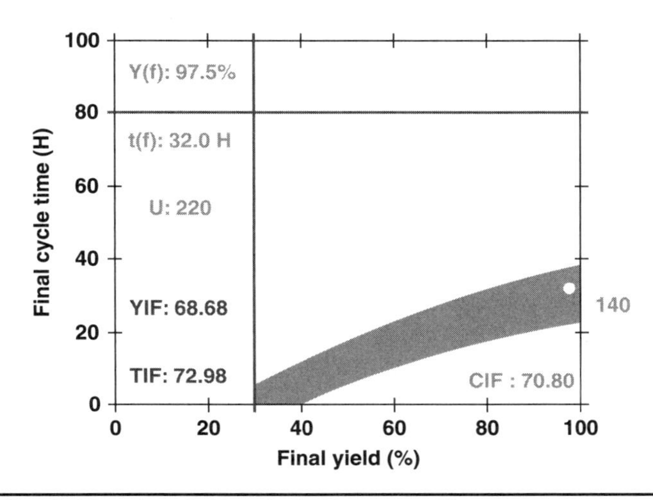

FIGURE D.8 Yield and cycle-time behavior—specific case.

calculated. In combination, they determine a CIF. The extreme cases for the situation shown are as follows:

Corner point of area	Yield	Time (h)
minimum yield, maximum time	30%	5.36
minimum yield, minimum time	39.5%	0.0
maximum yield, minimum time	100%	29.54
maximum yield, maximum time	100%	35.93

Any point within such limits will provide results such as subsequently shown.

D.4 Expected Production Output

How throughput improves over time can be expressed in two different ways. Using aggregate time vs. production output as the basis, the example produces Fig. D.9. It is quite obvious that production time for each unit is decreasing (i.e., the curve is flattening out up to 2.6 years) until the end of the "learning" period when it becomes constant. Fig. D.10 exhibits a bar chart of the yearly rework and output for this example. The reduction in rework, and increase

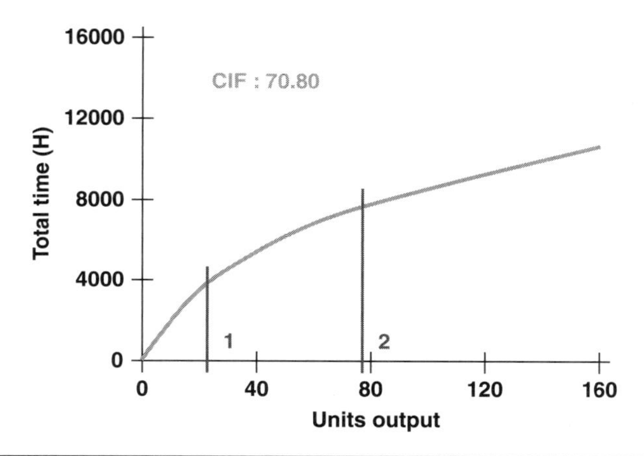

Figure D.9 Aggregate output characteristic—specific case.

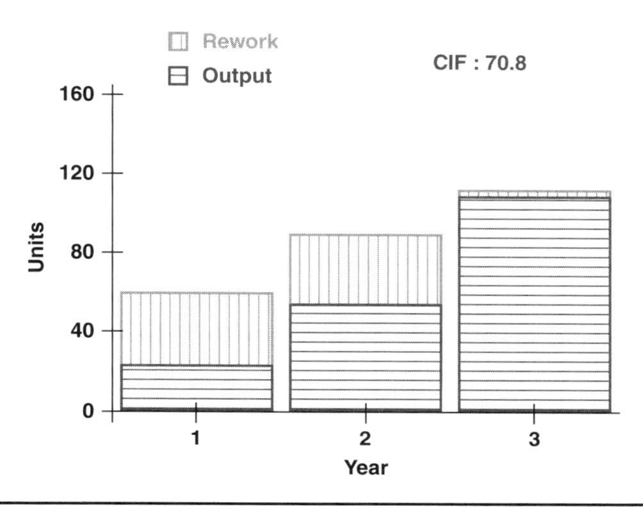

Figure D.10 Yearly output and rework behavior—specific case.

in output, for each year is readily apparent. For each of these output graphs, a table of the anticipated behavior is available:

Year	Units processed	Nominal time (h)	Final unit	Cycle-time (h)	Yearly output	Total output
1	59.70	64.32	59.70	52.67	22.72	22.72
2	90.28	42.53	149.99	35.51	54.37	77.10
3	112.02	34.28	262.00	32.00	109.22	186.31

D.5 Expected Costs

Each unit exhibiting the predicted behavior can also be evaluated for unit and total cost. A number of parameters need to be defined as seen in Fig. D.11. For some products, the material cost is significantly greater than any of the production costs. Labor cost is the most important for other products. Here, values for the example have been arbitrarily chosen such that those two costs for the final unit will be similar. If unit cost vs. output unit is plotted, the cost behavior can be

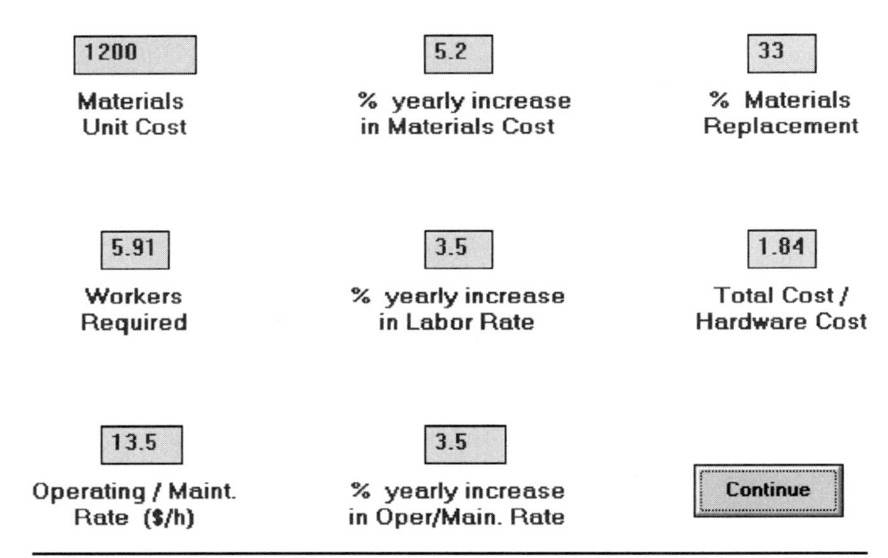

FIGURE D.11 Typical materials and labor parameters.

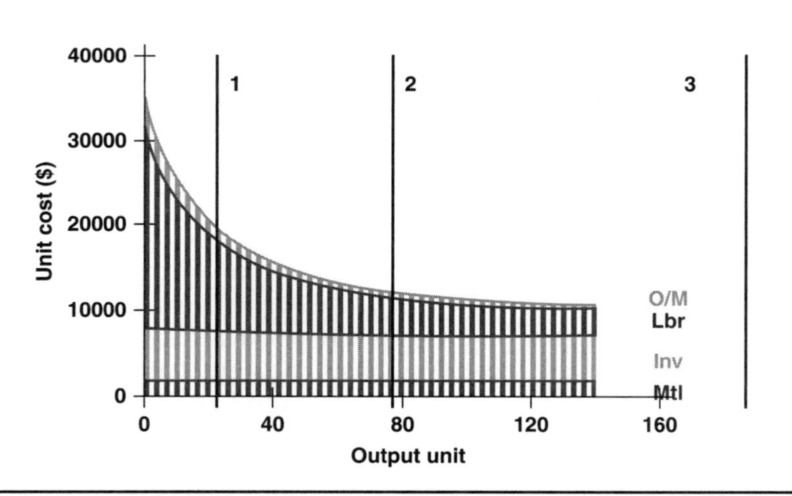

FIGURE D.12 Example cost vs. output characteristics.

seen in Fig. D.12. This data can also be expressed in tabular form as shown below:

Year	Units output	Units processed	Materials unit cost	Investment unit cost	Labor unit cost	Oper./main. unit cost	Total cost
1	22.7	59.7	1844	5751	14981	2281	564856
2	54.4	90.3	1538	5751	6480	987	802301
3	62.9	70.0	1378	5751	3502	533	702314
Total	**140**	**220**	**212168**	**805196**	**913062**	**139045**	**2069471**

Note that the investment cost is constant. As described in Chap. 1, it is the annualized cost (or sometimes called the capital recovery) for the example system.

D.6 Summary

Whether "start-up" behavior or a limited production run is to be estimated, the techniques shown in this appendix can be used. Simultaneous improvement in yield and reduction in cycle-time for each "good" unit produced can be readily predicted. The method has been used on a variety of products ranging from objects held in your hand to portions of giant physics experiments.

APPENDIX E

Two Case Study Summaries

E.1 Case Study Number 21—Automatic Transmission Final Assembly

This manufacturer wanted to evaluate final assembly of a new automatic transmission (for large vehicles) using automation as much as possible. Yearly production would be in the 50,000–250,000 range. Average wages plus benefits for a worker in that department were $21.25/h. Three shifts could be available for 240 work days per year. The minimum attractive rate-of-return was 16% over a 5-year capital recovery period.

Two generic system types were evaluated. The BASE system is mostly manual (MM); it would be similar to their current assembly methods. At the world-class (WC) level (with people, fixed automation, and programmable automation), the best ALTERNATIVE system had to be determined. Annual savings expected by using the WC system instead of the MM system for a range of production volumes are shown in Fig. E.1.

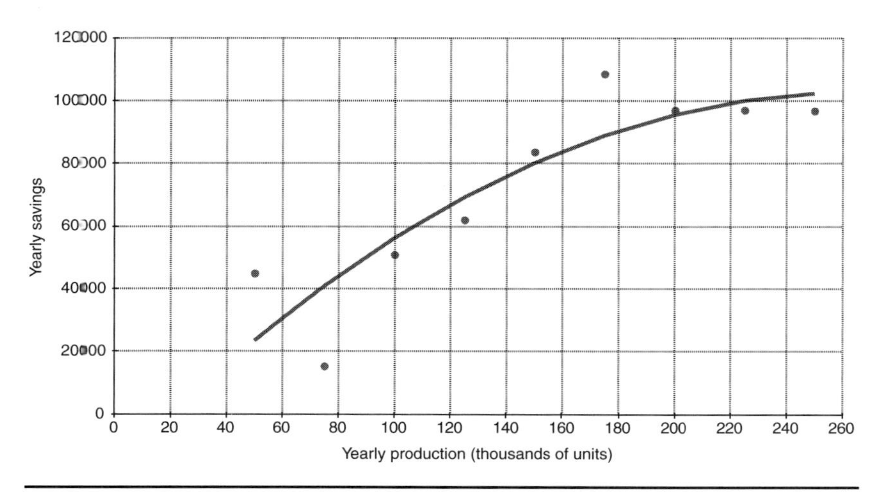

FIGURE E.1 Savings vs. yearly production—case study 21.

For the particular case of 160,000 units per year, the unit cost (using the MARR of 16%) reduces from $12.15 (MM) to $9.87 (WC). An apparent annual savings of $324,000 would occur. A more realistic way to financially compare these systems utilizes zero-present worth discounted cash flow (see App. A). The results show that the annual savings would be $84,000 (see Fig. E.1) with an actual IRoR of 42.3% and payback in 1.97 years.

Table E.1 summarizes the comparison between the WC and MM systems for this case study. The highest-level differentiation between them is as follows:

Basis for Comparison (product is . . .)	Current	New
Annual savings using alternative ($)	48,000	324,000
Actual internal rate-of-return (%)	14.76	42.33
Approximate break-even (years)	3.47	1.97

E.2 Case Study Number 24—Automatic Transmission and Differential Final Assembly

This manufacturer wanted to optimize final assembly of a new front wheel drive automatic transmission and differential assembly. Yearly production would be in the 400,000 to

Annual production quantity: **150,000;** loaded labor rate ($/h): **21.25** Working days per year: **240;** maximum shifts per day: **3** Capital recovery period: **5 years;** minimum attractive rate-of-return: **16%**		
Competing system specifications	**Base**	**Alternative**
Type	Mostly manual	World class
Manual stations (direct workers/shift)	10	4
Fixed automation stations	1	1
Robots (@ $80000)	0	2
Robots (@ $85000)	0	3
Robots (@ $110000)	0	0
Indirect workers/shift	2.3	2.4
Operating/maintenance rate ($/h)	6.2	11.5
New system cost ($) Capital equipment ($) Tooling ($)	1.01 million 0.47 million 0.49 million	2.24 million 0.94 million 0.54 million
Unit cost ($) Labor & oper./main. Capital recovery of investment	12.15 10.19 1.96	9.87 5.51 4.46

Table E.1 World-Class vs. Mostly Manual Comparison—Case Study 21

700,000 range. Average wages plus benefits for a worker in that department were $28.80/h. Two shifts could be available for 235 work days per year. The minimum attractive rate-of-return was 21% over a 10-year capital recovery period.

Two generic system types were evaluated. The BASE system is mostly manual (MM); it would be similar to their current assembly methods. At the world-class (WC) level (with people, fixed automation, and programmable automation), the best ALTERNATIVE system must be determined. Annual savings expected by using the WC system instead of the MM system for a range of production volumes are shown in Fig. E.2.

For the particular case of 550,000 units per year, the unit cost (using the MARR of 21%) reduces from $19.89 (MM) to $16.55 (WC). An apparent annual savings of $1.84 million would occur. A more realistic way to compare these systems

Annual production quantity: **550000;** loaded labor rate ($/h): **28.80** Working days per year: **235;** maximum shifts per day: **2** Capital recovery period: **10 years;** minimum attractive rate-of-return: **21%**		
Competing system specifications	**Base**	**Alternative**
Type	Mostly manual	World class
Manual stations (direct workers/shift)	71	35
Fixed automation stations	5	30
Automated pallet rotators	6	6
Robots (small SCARA type)	0	4
Robots (large SCARA type)	0	6
Indirect workers/shift	17	19
Operating/maintenance rate ($/h)	164	168
New system cost ($) Capital equipment ($) Tooling ($)	2.90 million 1.10 million 1.54 million	8.64 million 4.73 million 3.12 million
Unit cost ($) Labor & oper./main. Capital recovery of investment	19.89 18.15 1.74	16.55 11.44 5.11

TABLE E.2 World-Class vs. Mostly Manual Comparison—Case Study 24

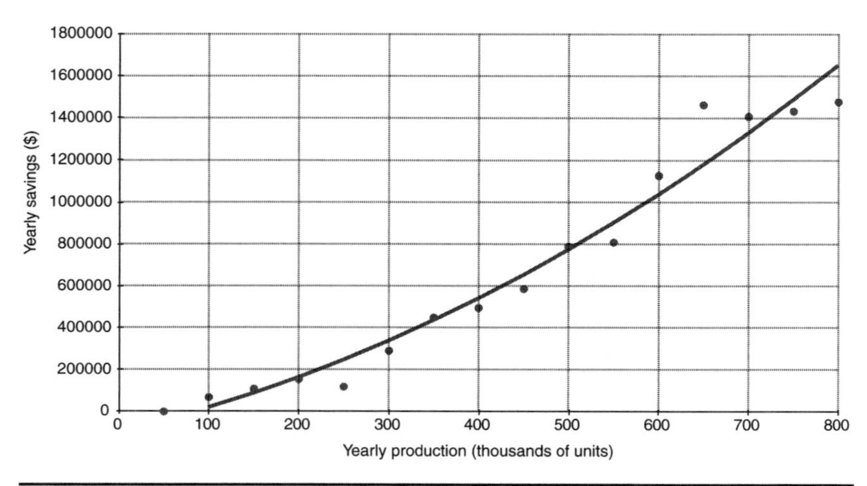

FIGURE E.2 Savings vs. yearly production—case study 24.

financially utilizes zero-present worth discounted cash flow (see App. A). Those results show that the annual savings would be $0.85 million (see Fig. E.2) with an actual IRoR of 42.7% and payback in 1.93 years.

Table E.2 summarizes the comparison between the WC and MM systems for this case study. The highest-level differentiation between them is as follows:

Basis for Comparison (product is . . .)	Current	New
Annual savings using alternative ($)	0.88 million	1.84 million
Actual internal rate-of-return (%)	24.59	42.67
Approximate break-even (years)	2.71	1.93

E.3 Summary

Although the economic parameters (minimum attractive rate-of-return, capital recovery period, and labor rate) used in these case studies are likely quite different from the values you will use, the methods needed to establish the most cost-effective system are exactly the same (see Chap. 7).

APPENDIX F

Advanced System Design Procedure

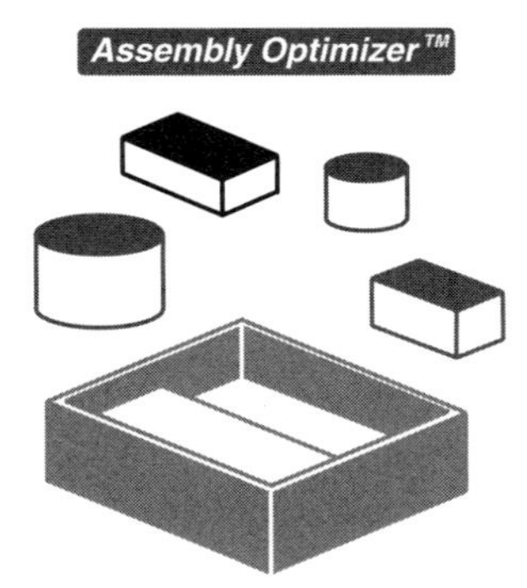

Advanced System Design Procedure

Demonstration Version

F.1 Introduction

The downloadable program (ASDP Demo) rapidly provides new insights into resource allocation problems. For the first time, systems designers have a robust, engineering tool that is not an operations research technique. These methods go beyond any analysis or simulation that you have previously done. Since the fundamental ideas can be applied to any situation in which a set of prescribed tasks has alternative methods for being accomplished, many other generically similar problems can be solved with often only slight changes to what you see here. The principal goal of the demonstration software is to design the "best" manufacturing systems. Assembly, fabrication, and/or material handling can be evaluated; emphasis will be on assembly.

The methods have been under development for more than two dozen years. They have been applied, at some stage during maturation, to a wide variety of products ranging

in **size** from a device that can be held in your hand to portions of Naval combatant ships,

in **production volume** from thousands of units per day to a single unit requiring multiple years,

in **process complexity** from 10 tasks to 98 tasks, and

in **resource types** from one to nine.

When manufacturing systems are to be designed using ASDP Demo, you will need information for:

tasks to be accomplished,

applicable resource types (people, fixed automation, robots, etc.), and

economic and production data.

Further guidance may be obtained from Gustavson (1984, 1996, 1997).

F.2 Basic Information

A reduced version of the system design program explained in this book along with example data files and a blank input data sheet at http://www.mhprofessional.com/getpage.php?c=engarch_updatezone_gustavson.php&cat=113. This ASDP Demo program contains the fundamentals for assembly (and other manufacturing) system design. You will need at least the following computer set-up:

Any PC or compatible with 1 MB RAM

Windows 3.1 or higher operating system

640 × 480 or higher resolution monitor

There is no INSTALL or SET-UP procedure required. The files may be copied to your hard disk; you should probably create a unique directory for them. To save altered data, be sure that files are changed from "read-only" to "archive."

F.3 Optimizing Assembly

One example has been provided for your edification. Fig. F.1 exhibits data for a reduced version of the air conditioning module final assembly used throughout this book; it maintains the essence of that electromechanical device. This example will allow you to exercise a number of fundamental features of the procedure. Various figures help guide you through the steps.

You will begin at the system design stage (tasks and applicable resource types have been determined). To be at that point in the design process, the following assumptions are made:

The product has been divided into nonseparable subassemblies (if necessary)

All of the available DfA and "ility" techniques have been implemented

Only one subassembly (or product) will be dealt with at a time

Company financial and work period data have been specified

The major phases of what will be achieved can be seen in the following diagram:

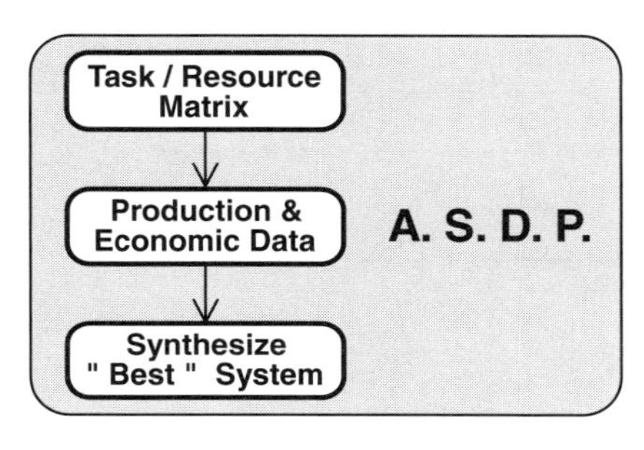

Applies to many types of manufacturing systems.

While the example is an assembly system (the production system design software is named A S D P—advanced system

Title A.C.M. Final Assy. Demo

ASDP

APPLICABLE TECHNOLOGY

240 Working days per year
3 Shifts available
5 s Station-to-station move time

14 % Min. Attractive Rate-of-Return
5 yr Capital Recovery Period
19 $/hr Ave. Loaded Labor Rate

RESOURCE data set name ACMRES-D

TASK data set name ACMTSK-D

Production Batch (units/yr)
175000

For each resource type :

C : Hardware cost ($)
ρ : Installed cost / hardware cost - rho factor
e : Up-time expected (%)
v : Operating / maintenance rate ($/hr)
t(c) : Tool change time (sec.)
m(s) : Maximum stations per worker
σ : Tooling cost / Support cost (%)
τ : Maximum tools at a station

When a resource can be used on a task :

Operation time (sec.) ; Tool number
Support hardware cost ($)

Number	RESOURCE TYPE / TASK	MNL	PS1	FXD			
		C: 1 ρ: 1.1 e: 87.5 v: .5 t(c): 1 m(s): .83 σ: τ: 6	C: 40000 ρ: 2.5 e: 85 v: .8 t(c): 4 m(s): 5 σ: τ: 4	C: 0 ρ: 1.5 e: 85 v: .8 t(c): 0 m(s): 4 σ: τ: 1	C: ρ: e: v: t(c): m(s): σ: τ:	C: ρ: e: v: t(c): m(s): σ: τ:	C: ρ: e: v: t(c): m(s): σ: τ:
1	Assemble (2), (3), (12), (13), (14), (15) & (16)	26 ; 101 1200	24 ; 201 20000	11 ; 302 170000			
2	Harness (10)	12 ; 104 500	15 ; 203 2000	X			
3	Evaporator Case (1)	10 ; 102 100	8 ; 202 10000	5 ; 304 60000			
4	Assemble (16), (5) & (6)	32 ; 105 3500	27 ; 204 13000	X			
5	Automatic Inspection	X	X	14 ; 307 100000			
6	Assemble (17), (18) & (19)	36 ; 108 7500	31 ; 206 15000	15 ; 309 150000			
7	Evaporator Core (4) 3 screws	12 ; 110 2000	10 ; 208 10000	X			
8	Core Shroud (7)-a 2 screws	10 ; 110 2000	8 ; 208 10000	X			
9	Heater core (7)-b Clamp (20), 3 screws	12 ; 110 2000	10 ; 208 10000	X			
10	Cover (9) 7 screws	31 ; 109 2100	22 ; 207 15000	9 ; 312 70000			
11	Final Inspection	15 ; 113 5000	X	12 ; 314 50000			
12	Pack	12 ; 114 100	10 ; 210 10000	X			
13							
14							

Prepared by ______ Date ______ Sheet 1 of 1

FIGURE F.1 Input data for limited ACM example.

design procedure), the techniques can be applied to flexible machining systems and material handling systems as well. A variety of other task/resource matrix problems can also be solved.

F.4 Design of Assembly Systems

The example will illustrate the basic characteristics and capabilities of ASDP. By removing some of the actual tasks, 12 (whose characteristics are defined) will remain. The requirements of those tasks will contribute important information needed for determination of the applicable resource types. One type is almost always manual; another type is almost always fixed automation. Since up to 3 resource types are allowed in this demonstration version of the program, you could then specify one programmable device (probably some type of robot).

You are now ready to tackle the fundamental goal; establish the most cost-effective combination of the applicable resource types to the tasks that must be performed using the ASDP Demo program. Ordering of the tasks *cannot* be altered within this program. Before beginning design of the assembly system, you must specify certain company financial parameters (e.g., wages plus benefits, minimum attractive rate-of-return, capital recovery period). When you want to enter your own data, use a format similar to the ASDP APPLICABLE TECHNOLOGY Data Sheet (see Fig. F.1 for the requirements).

Any solution found will satisfy all technological and economic requirements! It may be worthwhile to change the various parameters a couple times to see how the resulting set of systems gets altered. Any changed data will be retained. When starting the general solution process, you will need to decide the weighting factors to be used in the system RATING calculation. The method for ranking these solutions involves comparing three parameters to those for the maximum time solution:

Unit cost—combination of investment recovery, labor cost, and operating/maintenance costs

	Unit cost	33.34%
Weighting factors:	Investment	33.33%
	Stations	33.33%

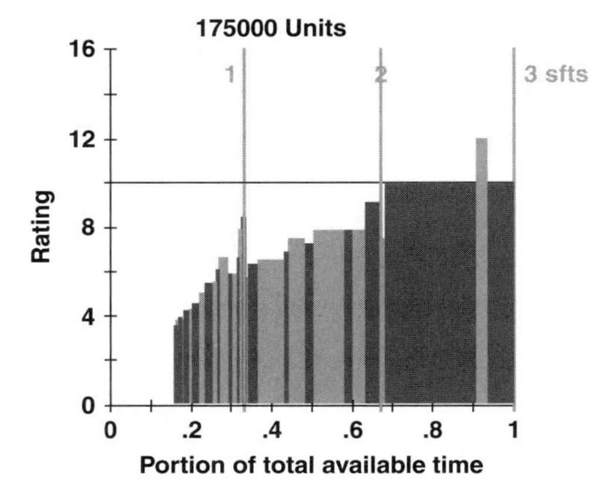

Avail. factor	Minimum RATING	pallets	Actual unit cost	Total (k) investment	Resources used: MNL	PS1	PA0
0.9332	11.93	3	2.054	533	1	2 ^	1
0.9034	10.01	4	2.338	633	1	3 ^	1
1.0000	10.00	4	1.948	762	0	3 ^	2
0.6666	9.05	5	2.196	760	1 ^	4	1
0.3332	8.44	9	2.386	576	5	3	1 ^
0.3176	7.86	10	2.638	988	6 ^	3	1

Figure F.2 Manufacturing view (general solution).

Investment required—hardware cost plus engineering, installation, debug, etc.

Number of workstations

The results for a particular production volume for the "reduced" ACM are shown in Fig. F.2, where you can see that 3 systems, in the 2 shift segment, have nearly the same rating (not a common occurrence). The upward arrow denotes which type of station is the bottleneck; such data is very important in some companies. Note that the highest-rated system does not necessarily occur at the maximum time

107.06 seconds Usable Cycle Time
85.00 % Bottleneck up-time 33.63 Units/hr expected
193688 units/yr Capacity of this System
532921 Total Investment, rho Factor = 2.07
257201 Capital Equipment 0 Tooling
1.85 Workers at 19.00/hr required
2.90/hr System Operating/Maintenance Rate
0.904 Year required for 3.0 Shift Operation
240 Days required for 2.71 Shift Operation

A.C.M. Final Assy. DEMO 175000 Units $2.054 Each 0.933 AF

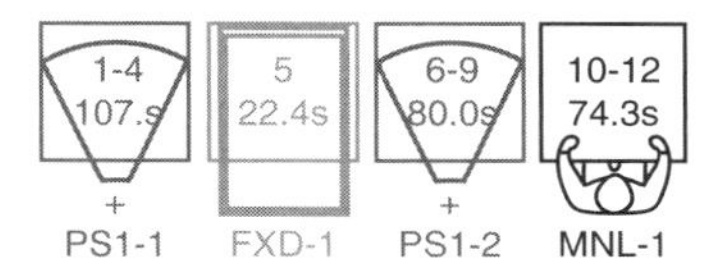

Figure F.3 Manufacturing view (best 3 shift solution).

condition (largest availability factor—at shift boundaries). Fig. F.3 exhibits the schematic layout for the best 3 shift solution. The bottleneck station for the system is readily seen. Fig. F.3 also displays various cost and performance characteristics:

Capacity—observe that the system can actually produce over 193,500 units per year (10% higher than needed)

Investment—shows total initial cost (which is annualized and divided by yearly production to establish part of the unit cost) as well as capital equipment and tooling costs

Workers—direct, if any, and indirect (part of unit cost)

Operating/maintenance cost for the system (part of unit cost)

Try some significantly different production volumes and/or alter the resource type uptimes to see what combinations of stations occur.

It is sometimes worthwhile to establish the sensitivity of unit cost vs. production volume above and below the most recently specified production quantity. Fig. F.4 exhibits the best 2 shift solution. When the production volume difference

74.29 seconds Usable Cycle Time
87.50 % Bottleneck up-time 48.46 Units/hr expected
279138 units/yr Capacity of this System
760011 Total Investment, rho Factor = 2.18
349101 Capital Equipment 0 Tooling
2.25 Workers at 19.00/hr required
4.50/hr System Operating/Maintenance Rate
0.627 Year required for 3.0 Shift Operation
240 Days required for 1.88 Shift Operation

A.C.M. Final Assy. DEMO 175000 Units $2.196 Each 0.667 AF

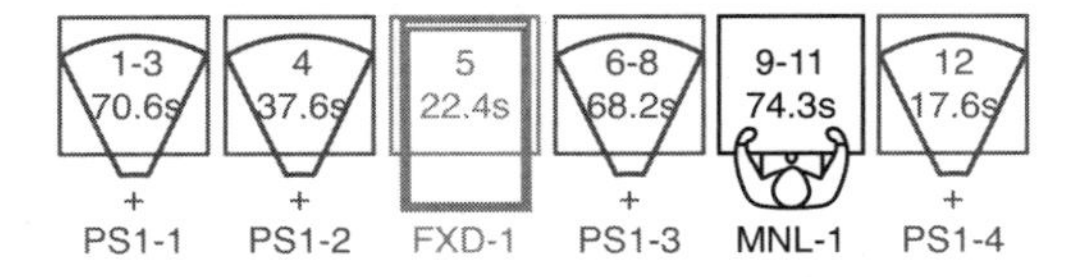

Figure F.4 Manufacturing view (best 2 shift solution).

is greater than 5% or so, the program should be re-run (this is an internal recycle) with that revised quantity since the best system could be quite different.

F.5 Limitations on Program

Data for a sample product has been provided. General methods for this solution procedure are readily evaluated. You may try your own data subject to limitations on the number of tasks and resource types.

Example products, or portions thereof, with a limited number of tasks (up to **12**) that need to be performed can be evaluated. When creating your own task/resource data, you can choose any resource types (**3** or less). Recall that cost and performance characteristics for each type of resource as well as for each applicable task must be specified.

References

Daschenko, A. I., "Optimization synthesis for building block assembly systems," *International Journal of Advanced Manufacturing Technology*, vol. 1, no. 5, Nov. 1986, pp 75–88.

Dorf, R. C., ed., "Economic justification," in *International Encyclopedia of Robotics: Applications and Automation*, Wiley, New York, 1988.

Graves, S. C., and D. E. Whitney, "A mathematical programming procedure for equipment selection and system evaluation in programmable assembly," in *Proceedings of the 1979 IEEE Decision & Control Conference*, Ft. Lauderdale, FL, Dec 1979.

Gustavson, R. E., "Choosing manufacturing systems based on unit cost," in *Proceedings of 13th ISIR/Robots 7*, Chicago, April 1983.

Gustavson, R. E., "Computer-aided synthesis of least cost assembly systems," in *Proceedings of 14th ISIR*, Gothenburg, Sweden, Oct 1984.

Gustavson, R. E., "S.P.M., a connection from product to assembly system," CSDL Report for DARPA available through NTIS, May 1990.

Gustavson, R. E., "Choosing an assembly system," *Assembly*, June/July 1996.

Gustavson, R. E., "Software guides assembly line designers," *Machine Design*, Oct 9, 1997.

Hodson, W. K., ed., *Maynard's Industrial Engineering Handbook*, McGraw-Hill, New York, 1992.

Holmes, C. A., "Equipment selection and task assignment for multiproduct assembly system," MIT Operations Research Center SM Thesis, Cambridge, MA, Jan 1987.

Kurtz, M., *Handbook of Engineering Economics*, McGraw-Hill, New York, 1984.

Lamar, B. W., and S. C. Graves, "An integer programming procedure for assembly system design problems," *Operations Research*, vol. 31, no. 3, May-June 1983, pp 522–545.

Nevins, J. L., and D. E. Whitney, ed., *Concurrent Design of Products and Processes*, McGraw-Hill, New York, 1989, Ch 12, 14, and 16.

Womack, J. P., and D. T. Jones, *Lean Thinking*, Simon & Schuster, New York, 1996, revised 2003.

Index

Note: Page numbers followed by *f* and *t* indicate figures and tables.

3-D view, 102–104

A

ACM Final Assembly, 138
activity-based costing, 6, 21, 41
actual unit cost, 50
additional tasks, 183
adhesive bond, 177
Advanced System Design Procedure (ASDP), 213–220
 basic information, 214–215
 Demo program, 214
 design of assembly systems with, 217–220
 input data, 216*f*
 limitations of demo, 220
 manufacturing view, 218–220*f*
 optimizing assembly with, 215–217
 overview, 213–214
 RATING weighting values for, 49*f*
 selection of minimum availability for, 48*f*
 system requirements, 214–215
aggregate output, 198
allowable investment, 149–158
 cash flow table, 151*t*, 153*t*
 fixed cost and, 150
 modified accelerated capital recovery system, 150
 overview, 148
 variable cost and, 150
 world-class system, 157–158
 zero present worth cash flow, 154*f*
analytical method, 5
annualized cost factor, 17, 20, 24, 25, 163
applicability of resource types, 38, 40*f*
applicable technology chart, 165–167
assembly process plan, 182–187
 altering data in, 184
 color coded, 183
 deleting tasks in, 184–185
 downward assembly, 182
 envelope size, 186
 fixed orientation, 183

assembly process plan (*Continued*)
  relative assembly difficulty, 186
  SCORE, 184–185
  steps in, 183
assembly sequence, 180–182
assembly system, 31–62
  base component, 178–179
  component/mate schematic for, 32–33
  details of, 53–58
  economic constraints, 41–45
  exploded view, 178, 179–180
  fixed automation category, 36
  fundamental principles, 32, 175–176
  hardware cost categories, 36
  in-process testing, 182
  management overview of, 58–59, 60–61*t*
  manual labor category, 36
  optimizing, 215–217
  process plan for, 33–41, 35*t*
  production requirements, 40*f*, 41–45
  resource types, 53–58
  sequence, 180–182
  spectrum for range of production volume, 62, 116*f*
  task data for, 173–187
    assembly process plan, 182–187
    assembly sequence, 180–182
    base component, 178–179
    component data, 176
    exploded view, 179–180
    exploded view of assembly, 178
    fundamental principles, 175–176
    in-process testing, 182
    input data requirements, 176–178
    mate data, 176–178
  tasks, 33
  usable systems in, 47–53
automatic transmission final assembly, 205–209. *See also* electromechanical products
  ALTERNATIVE system, 205
  BASE system, 205
  case studies, 205–209
  mostly manual vs. world-class system, 205–206, 207–208*t*
  savings vs. yearly production, 206*f*, 208*f*
available time, 18–19, 24, 43*f*, 53, 117*f*, 218*f*
available time (AVLTIM), maximum, 24
average loaded labor rate, 43
average workstation time, 82

B

base component, 178–179
BASE system, 205, 207
batch-of-one, 9
bearing race, 177
best-case random number behavior, 72, 89
bolted joints, 177
bottleneck station, 89–90
  asynchronous systems, 66
  degree of difficulty and, 122*t*
  manufacturing view, 219*f*
  minimum time, 89
  uptime expected, 56
bottom-up method, 161
bounding box dimensions, 176
bushings, 177

C

capacity, 219
capital equipment cost, 59, 208*t*
capital equipment depreciation, 150
capital recovery, 24, 43, 208*t*
capital recovery factor, 163
capital recovery period, 43, 138, 207*t*
cash flow, projected, 151*t*
cash flow table, 153*t*
Charles Stark Draper Laboratory, 10
closed loop system, 96–101
  spaced loop, 100*f*, 101*f*
  with spacing, 96, 98–101
  tight loop, 99*f*, 100*f*
  without spacing, 96, 97–98
cocure, 177
code for component, 176
collaborative engineering, 10
color coded tasks, 183
combined improvement factor (CIF), 194, 198
component codes, 177
component data, 176
component schematic, 32–33, 34*f*
component tasks, 183
component/mate schematic, 32–33, 34*f*
consecutive tasks, 45
constant value situation, 138–141
continuous improvement, 9, 162
cost and utilization behavior, 59*f*
cost assignments, 26*t*
cost comparison equation, 163–165
cost improvement, 136, 139, 140*f*, 142–145*t*
cost-effective system, 162
costs, expected, 199–200
critical alignment, 177
cumulative distribution function (CDF), 70–71, 71*f*, 81*f*
cycle time, 191–201
  asynchronous systems, 66
  expected costs and, 199–201
  expected production output and, 198–199
  improvement behavior, 193*f*
  production improvement and, 194–198
  unit cost and, 164
  vs. unit produced, 192*f*
  yield and, 196*f*, 197*f*
cycle time expected, 56
cycle time reduction, 194

D

degree of difficulty, 33, 38, 179, 183–184, 186
depreciable portion, 137, 150
depreciation method, 150
differential and automatic transmission final assembly, 206–209
direct labor factor, 53
direct workers, 219
discounted net income, 150, 155
discrete event distributions, 66–68
downward assembly, 182–187

E

economic constraints, 32, 41–45, 47, 110
economic data, 43*f*, 196*f*
economic justification, 6–7, 136
economic requirements, 17
economic-technological synthesis of systems, 161–169
  annualized cost factor, 163

economic-technological synthesis of systems (*Continued*)
  applicable technology chart, 165–167
  bottom-up method, 161
  capital recovery factor, 163
  cost comparison equation, 163–165
  least-cost system, 167–169
  top-down method, 161
  utilization, 165
effective capacity, 59
electromechanical products. *See also* automatic transmission final assembly
  choice of resource types for, 40*f*
  component/mate schematic for, 34*f*
  cost and utilization behavior, 59*f*
  general solution for, 52*f*
  management view for assembly of, 60–61*f*
  portion of assembly plan for, 35*t*
  portion of specific solution for, 54*t*, 57*f*
  primary factors for, 37*f*
  resource type data, 26*t*
  specific solution for, 54*t*, 57*f*
  task/resource type matrix, 42*f*
  unit cost and total investment vs. batch size for, 62*f*
engineering economy, 7, 149
envelope size, 186
expected value for station time, 69
exploded view, 178, 179–180
exponential distribution, 78–87
  actual value, 79
  application of, 81*f*
  application to system synthesis, 89–92
  behavior, 68*f*
  cumulative distribution function for, 81*f*
  definition of, 67
  equation, 80
  expected value, 79
  inverse cumulative distribution function for, 82*f*
  maximum station time, 79
  minimum station time, 79
  stochastic results, 83–85*t*
  stochastic station times, 86*f*
  stochastic time ratios, 87*f*

F

final cycle time, 196–197*f*, 197
final year % salvage value, 138
final yield, 196*f*, 197, 197*f*
fixed automation category, 36
fixed automation stations, 57, 96, 207*t*, 208*t*
fixed cost, 150, 163
fixed orientation, 183
fixed unit cost, 56
flexibility, 24
flexibility factor, 15, 19–20, 24
force (or load) required, 33

G

gears, 177
general placement, 177
geometric layouts, 95–96

H

hardware cost, 36, 38
hardware price, 21
heuristic method, 16
hourly wage, 138

I

incremental depreciable portion, 137
incremental hardware cost, 136–137
incremental major equipment cost, 136
incremental minor equipment cost, 136
incremental system cost, 136
indirect workers, 208*t*, 219
industrial robots, 3–4
  economic justification, 6–7
  internal organization and, 4–5
  mostly manual vs. world-class system, 208*t*
in-parallel stations, 44, 47, 111, 115
in-parallel stations, maximum, 44
in-process testing, 182
institutional costs, 149
intermediate (in-process) tests, 182
internal organization, 4–5
internal rate-of-return (IRoR), 7, 136, 141*f*, 150, 151*t*, 153*t*, 158, 206*t*, 209*t*
intervening tasks, 45*f*
inverted yield improvement behavior, 195*f*
investment required, 47, 56, 120*t*, 157*f*, 163, 218
investments, 219
isometric view, 102–104

J

just-in-time, 9

K

kan-ban, 9

L

labor cost, 200*f*, 208*t*
labor rate, 43
lean production, 6
learning curves, 191
learning rate, 191
least-cost system, 167–169
line balance, 32, 58
linear system layout, 95, 97
load required, 33
loaded labor rate (hourly wages + benefits), 17, 21, 43, 150

M

maintenance rate, 43
manual labor category, 36
manual resource type, 57
manual stations, 57, 96, 207*t*, 208*t*
manufacturing cells, 9
manufacturing methodologies, 7–10
manufacturing system design, 4–5
  economic justification and, 6–7
  internal organization and, 4–5
  methodologies, 7–10
  overview, 3–4
Massachusetts Institute of Technology, 16
mate data, 176
mate schematic, 32–33, 34*f*
material cost, 149
materials, 200*f*
maximum available time (AVLTIM), 24
maximum in-parallel stations, 44
maximum number of tasks (NTSKS), 24
maximum shifts available, 44

maximum station time, 19, 69, 79, 97
maximum stations per worker, 17
maximum tools, 44*f*, 114
method time measurement (MTM), 37
minimum attractive rate of return (MARR), 7, 43, 138–139, 144*f*, 206
minimum pallets, 50, 55*t*, 117*f*, 169*t*
modified accelerated capital recovery system (MACRS), 150
mostly manual (MM) system, 135–145. *See also* world-class (WC) system
  annual cost vs. investment, 156*f*
  case studies, 205–206, 207–209
  changes in yearly costs, 143–145
  changes in yearly production volume, 142–143
  constant value situation, 138–141
  cost and performance characteristics, 137*t*
  definition of, 136
  incremental major/minor equipment cost, 136
  incremental system cost, 136
  nonconstant yearly costs, 141
  unit cost improvement, 140*f*
  vs. world-class system investment, 158*f*
  yearly savings characteristics, 139*f*
motions required, 33
multiple-product system, 109–131
  cost and performance data, 119*t*
  cost and performance summary, 120*t*
  details of, 118–123
  fundamental principles, 110–111
  group of usable systems, 114–118
  labor cost, 118
  management overview of, 123–131
  management view of the best system, 124–129*t*
  maximum tools at a station, 114
  overview, 109–110
  production requirements, 111–114
    maximum tools at a station, 114
    production batch size, 111
  RATING and ranking, 117*f*
  resource cost, 118
  schematic layout, 121–123*f*
  summary effort at workstations, 130*f*
  task/resource matrix, 110–111
  unit cost vs. yearly production vs. investment, 116*f*
  workstation time allocation, 113*f*

N

National Science Foundation, 10
net income, 150
net profit, 155*f*
new equipment, 45*f*
new system cost, 208*t*
nonconstant yearly costs, 141

number of different tools (NTOOL), 24
number of tasks (NTSKS), maximum, 24

O

operating rate, 43
operating/maintenance costs, 219
operating/maintenance factor, 53
operating/maintenance rate, 17, 208*t*
operation time, 19–20, 54
operations research (OR) methods, 16
optimal solution, 15
organizational chart, 5

P

parabolic output, 144–145*t*
parabolic production, 143
parallel stations, 58
payback period, 155
planning horizon, 141*f*, 144*f*, 151
principal assembly direction, 175, 176, 178
probability density function, 69
probability distribution function (PDF), 70
process plan, 33–41, 182–187
process time, 38
product design, 5, 9–10, 174
production batch size, 44, 111–112, 174
production capacity, 56
production improvement, 194–198
production improvement behavior (PIB), 194
production mix, 112
production output, expected, 198–199
production requirements, 41–45, 88*f*
production volume. *See also* assembly system
  advanced system design procedure and, 214
  annual capacity, 75
  assembly system design and, 218–220
  batch size, 112
  constant value situation and, 138
  costs and, 123
  flexibility factor and, 20
  vs. general cost, 18*f*
  spectrum of systems for, 62, 116*f*
  task/resource data and, 41
  time requirements, 17
  yearly, changes in, 142–143
programmable automation, 15, 21, 36–37, 57
programmable automation stations, 96
projected cash flow, 153*t*
pull method, 9, 66
push method, 8, 65

Q

quality rating, 16, 22

R

random number behavior, 89–90
random number generator, 70
random number seed (RNS), 89–90
  common value, 72
  cumulative distribution function and, 70
  unique value, 72
RATING weighting, 47–50, 115
relative assembly difficulty, 186
replication number, 19

representative costs, 23*t*
residual loan, 163
residual value, 43
resource, 55
resource cost, 53
resource flexibility, 15, 19–20, 24
resource type data, 25*t*
resource types, 17
  applicability of, 38, 40*f*
  cost, 17
  costing, 38, 39*t*
  programmable, data for, 41*f*
  for typical electromechanical products, 40*f*
resources used, 51, 53
rho factor, 7, 17, 54*t*, 119*t*, 121–123*f*, 150
rivets, 177
robots, 3–4
  economic justification, 6–7
  internal organization and, 4–5
  mostly manual vs. world-class system, 208*t*

S

salvage value, 137*t*, 138
savings, vs. yearly production, 208*f*
schematic layout, 96–97
selective fit, 177
shifts, 138
shifts available, 17, 25*t*, 44, 138
simulation, 8–9, 65, 72
Sloan School of Management, 10, 16
solder, 177
spaced loop system, 103*f*
special equipment cost, 17
special equipment identifier, 17
splines, 177
start-up phase, 191
station cost, 36, 44
station time, maximum, 19, 69, 79, 97
stations per worker, 17, 19, 21, 42*t*
station-to-station move time, 17, 27, 44, 54, 138
statistical process control, 9
steady-state condition, 193
stochastic analysis, 65–92
  application to manufacturing system, 71–78
  application to synthesis of systems, 89–92
  discrete event distributions, 66–68
  exponential distribution, 78–87
  overview, 65–66
  triangular distribution, 68–71, 73–74*t*, 76*t*
stochastic time, 78*f*, 82, 87*f*, 90
support cost, 55
symbolic name, 17
system annualized charge factor, 56
system configurations, 95–105. *See also* assembly system
  3-D view, 102–104
  closed loop (with spacing), 98–101
  closed loop (without spacing), 97–98
  geometric layouts, 95–96
  linear system layout, 97
  schematic layout, 96–97
  "U" cell system, 101–102
system cost, 163–164
system design, 16–18
  allocation of time used, 19
  available time for resource, 18–19
  basics of, 16–18
    economic requirements, 17
    resource types, 17

tasks, 17
time requirements, 17
fixed cost of station, 20–21
fundamental principles, 32
input data, 24
overview, 15–16
quality rating, 22
resource flexibility, 19–20
solution procedure, 22–24
variable cost for task, 21–22
system operating/maintenance rate, 56

T

task actions, 33
task data for assembly systems, 173–187
assembly process plan, 182–187
assembly sequence, 180–182
base component, 178–179
component data, 176
exploded view, 179–180
exploded view of assembly, 178
fundamental principles, 175–176
in-process testing, 182
input data requirements, 176–178
mate data, 176–178
overview, 173–174
task degree of difficulty, 33, 38, 179, 183–184, 186
task description, 33, 35*t*, 42*t*, 59, 183, 186
task number, 40*f*, 53
task time, 17, 20, 26*t*, 33, 37, 54, 112, 131
task type, 35*t*, 178, 186
task/resource matrix, 33, 41, 42*t*, 110–111
tasks, 17
number of, 56
variable cost, 21–22
time constraints, 138, 139, 167
time data, 196*f*
time improvement factor (TIF), 193
time reduction, 193
time requirements, 17
time used, 55–56
tool change, 54
tool change time, 18, 20, 27
tool hardware cost, 38
tool number, 27, 38, 55, 118
tooling cost, 44*f*, 59, 208*t*
tooling depreciation, 150
tools, maximum, 44*f*, 114
tools, number of, 56
top-down method, 161
torqued fasteners, 177
total hardware cost, 56
total investment, 50
total investment required, 56
triangular distribution, 67*f*, 68–71
definition of, 67
stochastic production ratios, 78*f*
stochastic results, 73*t*, 76*t*
stochastic station times, 77*f*
synthesis of systems, 89–91
workstation throughput time, 77*f*

U

"U" cell system, 101–102
definition of, 96
isometric view of, 104*f*
perspective view of, 105*f*
schematic layout, 103*f*
ultrasonic weld, 177
uniform distribution, 67, 67*f*
unit cost, 47, 50, 56, 164, 208*t*, 217

unit cost improvement, 139, 140*f*, 142–145*t*
units per hour, 56
units per pallet, 44, 45*f*
units processed, 192*f*
up-time expected, 166*f*, 216*f*
usable systems, 47–53. *See also* assembly system
  availability factor, 50
  investment required, 47
  minimum pallets, 50
  multiple-product system, 114–118
  parameters specified, 48
  RATING weighting values for, 48
  resources used, 51–53
  specific values not used, 48
  total investment, 50
  unit cost, 47
used equipment, 45*f*
utilization, 165
utilization behavior, 59*f*

V

variable cost, 53, 150, 163
variable unit cost, 56

W

weight, 176
weighting factors, 47–49, 51*f*, 115, 217, 218*f*
workers, number of, 56, 219
working days per year, 17, 44, 138, 207*t*
workstations. *See also* assembly system
  capital equipment cost, 59
  effective capacity, 59
  expected cycle time, 58
  with longest time, 96
  with maximum required effort, 96
  number of, 218
  resource type, 58, 59
  station number, 59
  task description, 59
  tasks, 58
  throughput time, 77*f*
  tooling cost, 59
world-class (WC) system, 135–145. *See also* mostly manual (MM) system
  allowable investment, 157–158
  annual cost vs. investment, 156*f*
  case studies, 205–206, 207–209
  changes in yearly costs, 142–143
  changes in yearly production volume, 142–143
  constant value situation, 138–141
  cost and performance characteristics, 137*t*
  definition of, 136
  incremental major/minor equipment cost, 136
  incremental system cost, 136
  vs. mostly manual savings, 139*f*
  vs. mostly manual system investment, 158*f*
  nonconstant yearly costs, 141
  unit cost improvement, 140*f*
worst-case random number behavior, 72, 90

Y

yearly production volume, 142–143
yearly production, vs. savings, 208*f*

yield, 191–201
  cost vs. output
    characteristics, 200*f*
  cycle time and, 196*f*, 197*f*
  expected costs, 199–201
  expected production output,
    198–199
  materials/labors parameters,
    200*f*
  production improvement,
    194–198
yield improvement factor
  (YIF), 194

Z

zero present worth cash flow,
  152*f*, 154*f*, 155